CATALYSIS LOOKS TO THE FUTURE

Panel on New Directions in Catalytic Science and Technology

Board on Chemical Sciences and Technology

National Research Council

NATIONAL ACADEMY PRESS
Washington, D.C. 1992

Support for this project was provided by the U.S. Department of Energy under Grant No. DE-FG05-90ER14103; the National Science Foundation under Grant No. CTS-8921829 and CHE-8921664; Air Products and Chemicals, Inc.; Catalytica, Inc.; Chevron Research Company; Dow Chemical USA; Exxon Research and Engineering Company; E. I. Du Pont de Nemours and Company, Incorporated; Mobil Research and Development Corporation; UOP Inc.; and the National Academy of Engineering.

Library of Congress Catalog Card Number 91-66333
International Standard Book Number 0-309-04584-3

Additional copies of this report are available from:

National Academy Press
2101 Constitution Avenue, NW
Washington, DC 20418

S447

Printed in the United States of America

Contents

CATALYSIS LOOKS TO THE FUTURE

Executive Summary

SOCIETAL IMPACT OF CATALYTIC SCIENCE AND TECHNOLOGY

The chemical industry is one of the largest of all U.S. industries, with sales in 1990 of $292 billion and employment of 1.1 million.[1] It is one of the nation's few industries that produces a favorable trade balance; the United States now exports chemical products amounting to almost twice the value of those that it imports (exports of roughly $37 billion compared to imports valued at about $21 billion).[2] Between 1930 and the early 1980s, 63 major new products and 34 process innovations were introduced by the chemical industry. More than 60% of the products and 90% of the processes were based on catalysis. Catalysis also lies at the heart of the petroleum refining industry, which had sales in 1990 of $140 billion and employed 0.75 million workers.[3] Clearly then, *catalysis is critical to two of the largest industries in sales in the United States; catalysis is also a vital component of a number of the national critical technologies identified recently by the National Critical Technologies Panel.[4]*

Looking into the future, one can see many exciting challenges and opportunities for developing totally new catalytic technologies and for further

---

[1]U.S. Department of Commerce, *U.S. Industrial Outlook 1991*, International Trade Administration, Washington, D.C., 1991.

[2]U.S. Department of Commerce, *U.S. Industrial Outlook 1991*.

[3]U.S. Department of Commerce, *U.S. Industrial Outlook 1991*.

[4]*Report of the National Critical Technologies Panel*, William D. Phillips, chair, Arlington, Va., March 1991.

improving existing ones. Increasing public concern with the effects of chemicals and industrial emissions on the environment calls for the discovery and development of processes that eliminate, or at least minimize, the use and possible release of hazardous materials. Concern with the environment and the supply of raw materials is also focusing attention on opportunities for recycling. Of particular interest for the chemical industry is the prospect of producing polymers that are readily recyclable. Although the world supply of petroleum is adequate for current demand, there is a need to continue the search for technologies that will permit the conversion of methane, shale, and coal into liquid fuels at an acceptable cost. Also, to maintain their economic competitiveness, U.S. producers of commodity and fine chemicals will need to shift to lower-cost feedstocks and processes exhibiting higher product selectivity. Taken together, these forces provide a strong incentive for increasing research efforts aimed at the discovery and development of novel catalytic processes and for continuing to extend the frontiers of catalytic science.

The following are benchmark discoveries made over the years in the science and technology of catalysis:

• 100 years ago: Paul Sabatier (Nobel Prize 1912) at the University of Toulouse started work on his method of hydrogenating organic molecules in the presence of metallic powders.

• 70 years ago: Irving Langmuir (Nobel Prize 1932) at General Electric laid down the scientific foundations for the oxidation of carbon monoxide.

• 50 years ago: Vladimir Ipatieff and Herman Pines at UOP developed a process to make high-octane gasoline that was shipped just in time to secure the victory of the Royal Air Force in the Battle of Britain.

• 30 years ago: Karl Ziegler and Giulio Natta (Nobel Prize 1963) invented processes to make new plastic and fiber materials.

• 17 years ago: W. S. Knowles at the Monsanto Company obtained a patent for a better way to make the drug L-Dopa to treat Parkinson's disease.

• Today: Thomas Cech (Nobel Prize 1989) at the University of Colorado received U.S. patent 4,987,071 to make ribozymes, a genetic material that may, one day, be used to deactivate deadly viruses.

The above examples deal with materials for health, clothing, consumer products, fuels, and protection of the environment, but all have a common feature: they rely on chemical or biochemical catalysts.

WHAT ARE CATALYSTS?

What are catalysts, these substances that hold the keys to better products and processes, and continue to have a strong impact on our health, economy, and quality of life? A catalyst is a substance that transforms reactants into products, through an uninterrupted and repeated cycle of elementary steps, until the last step in the cycle regenerates the catalyst in its original form. More simply put, a catalyst is a substance that speeds up a chemical reaction without itself being consumed in the process. Many types of materials can serve as catalysts. These include metals, metal compounds (e.g., metal oxides, sulfides, nitrides), organometallic complexes, and enzymes.

CATALYTIC SCIENCE AND TECHNOLOGY

The first triumph of a large-scale catalytic technology goes back to 1913 when the first industrial plant to synthesize ammonia (NH_3) from its constituents, elemental nitrogen (N_2) and elemental hydrogen (H_2), was inaugurated in Germany. From the outset, and until the present, the catalyst in such plants has consisted essentially of iron. The mechanism of the reaction is now well understood. Certain groups of iron atoms at the surface of the catalyst are capable of dissociating first a molecule of nitrogen and then a molecule of hydrogen, and finally of recombining the fragments to ultimately form a molecule of ammonia. The catalyst operates at high temperature to increase the speed of the catalytic cycle and at high pressure to increase the thermodynamic yield of ammonia. Under these conditions, the catalytic cycle turns over more than a billion times at each catalytic site, before the catalyst has to be replaced. This high productivity of the catalyst explains its low cost: the catalyst results in products worth 2000 times its own value during its useful life.

The next illustration of catalysis shows that industrial catalysts can be biomimetic in the sense that they imitate the ability of naturally occurring enzymes to produce optically active molecules. Many pharmaceuticals are known to be active in only one form, let us say the left-handed form. It is therefore critical to obtain the left-handed form with high purity. It is particularly important when the drug is toxic, even if only slightly so, and must be administered over many years. It is true of a molecule called L-Dopa used in the treatment of Parkinson's disease. The right-handed molecule is inactive. In ordinary synthesis, both forms are produced in equal amounts. Their separation is costly. Is it possible to produce only the left-handed form by means of a synthetic catalyst? The first successful industrial synthesis of this kind was achieved by Monsanto, and a patent for the selective synthesis of L-Dopa was granted in 1974. The catalytic process used to make L-Dopa today is an important achievement in industrial catalysis.

Finally, recent and current developments in catalytic technology are targeted at the protection of the environment. The best-known example deals with catalytic converters that remove pollutants from the exhaust gases of automobiles. Catalytic converters for automobiles were first installed in the United States in the early 1970s. These devices were subsequently introduced in Japan and are currently spreading through the European Community and Switzerland. The most advanced catalyst now contains three metals of the platinum group and controls the emission of carbon monoxide, nitrogen oxides, and unburned gasoline molecules by use of a very complex network of catalytic reactions. This application has contributed more than any other to public awareness of catalysis and of its many applications for the benefit of mankind.

RESEARCH OPPORTUNITIES IN CATALYTIC SCIENCE AND TECHNOLOGY

For viable commercial application, catalysts of any type—heterogeneous, homogeneous, or enzymatic—must exhibit a number of properties, the principal ones being high activity, selectivity, and durability and, in most cases, regenerability. The activity of a catalyst influences the size of the reactor required to achieve a given level of conversion of reactants, as well as the amount of catalyst required. The higher the catalyst activity, the smaller are the reactor size and the inventory of catalyst and, hence, the lower are the capital and operating costs. High catalyst activity can also permit less severe operating conditions (e.g., temperature and pressure), and this too can result in savings in capital and operating costs. The amount of reactant required to produce a unit of product, the properties of the product, and the amount of energy required to separate the desired product from reactants and by-products are all governed by catalyst selectivity. As a consequence, catalyst selectivity strongly influences the economics of a process. Catalyst productivity and the time on-stream are dictated by catalyst stability. All catalysts undergo a progressive loss in activity and/or selectivity with time due to chemical poisoning, denaturing, thermal deactivation or decomposition, and physical fouling. When the decrease in performance becomes too severe, the catalyst must be either regenerated or replaced. In view of this, high stability and ease of regeneration become important properties.

Catalysis is a complex, interdisciplinary science. Therefore, progress toward a substantially improved vision of the chemistry and its practical application depends on parallel advances in several fields, most likely including the synthesis of new catalytic materials and understanding the path of catalytic reactions. For this reason, future research strategies should be focused on developing methods with an ability to observe the catalytic reaction steps in situ or at least the catalytic site at atomic resolution. There is also a need to link heterogeneous catalytic phenomena to the broader knowledge base in solutions and in well-defined metal complexes.

Substantial progress and scientific breakthroughs have been made in recent years in several fields, including atomic resolution of metal surfaces, in situ observation of an olefin complexed to zeolite acid sites by nuclear magnetic resonance (NMR) spectroscopy, and in situ characterization of several reaction intermediates by a variety of spectroscopic techniques. Theoretical modeling is now ready for substantial growth as a result of progress in computer technology and in theory itself. For these reasons, it is desirable to focus on areas in which the extensive scientific and technological resources of academe and industry may lead to the fastest practical results. In order of priority, these areas are

1. in situ studies of catalytic reactions;
2. characterization of catalytic sites (of actual catalysts) at atomic resolution (metals, oxides);
3. synthesis of new materials that might serve as catalysts or catalyst supports; and
4. theoretical modeling linked to experimental verification.

Furthermore, additional steps must be taken to facilitate interaction and, in fact, cooperation between industry, dealing with proprietary catalysts, and academe, developing advanced characterization tools for catalysis.

REPORT FINDINGS

Catalysis plays a fundamental role in the economy, environment, and the public health of the nation; it underlies several of the critical national technologies identified in the 1991 *Report of the National Critical Technologies Panel*. Specifically, two of the largest industry segments, chemicals and petroleum processing, depend on catalysis; many of the modern, cost- and energy-efficient environmental technologies are catalytic; and biocatalysis offers exciting opportunities for producing a broad range of pharmaceuticals and specialty chemicals, and for bioremediation of the environment.

As opposed to some other areas of technology, the United States still plays a leadership role in catalysis, as evidenced by the general superiority of U.S petroleum conversion processes and most chemical processes, and by the net positive chemical trade balance of the United States. However, this position is eroding rapidly, due to the heavy investment in R&D of Japan and the European Community.

Major opportunities exist for developing new catalytic processes and for advancing the intellectual frontiers of all branches of science pertaining to catalysis. *To take advantage of these opportunities, careful attention must be given to effective use of the nation's resources, so that the United States can maintain its leadership role.*

REPORT RECOMMENDATIONS

For Industry

Industry in the United States is faced with enormously high costs associated with environmental remediation, short-term financial performance, high interest rates, and global competition. At the same time, substantial opportunities exist for developing new processes and products based on still-to-be-discovered catalysts. Therefore, industry should strive for an updated balance between long- and short-range research, aimed at taking advantage of these opportunities. This would be facilitated by long-range business and technology planning; technology forecasting and trend analysis; a more stable commitment to strategic projects, joint development, and joint venture programs with other companies for risk sharing; and high-quality project selection and evaluation methodologies. The challenges faced by industry will require additional advances in the science of catalysis, as well as advances in instrumentation. To better achieve this goal, additional opportunities for developing meaningful collaborative programs in partnership with academic and national laboratory researchers should be pursued.

Two elements are recommended as essential:

1. Enhanced appreciation by academic researchers of industrial technology. Vehicles for this include
- **long-term consulting arrangements involving regular interactions with industrial researchers,**
- **sabbaticals for industrial scientists in academic or government laboratories,**
- **sabbaticals for academic or government scientists in industrial laboratories,**
- **industrial internships for students,**
- **industrial postdoctoral programs, and**
- **jointly organized symposia on topics of industrial interest.**

2. Increased industrial support of research at universities and national laboratories. Vehicles for this include
- **research grants and contracts;**
- **unrestricted grants for support of new, high-risk initiatives; and**
- **leveraged funding (e.g., support of the Presidential Young Investigators program.)**

For Academic Researchers

Academic researchers have made major contributions to the fruitful understanding of the structure of homogeneous, heterogeneous, and enzyme catalysts, and the relationships between structure and function. They have also contributed substantially to the development of new instrumental and theoretical techniques, many of which now find application in industrial laboratories. To maintain healthy progress in catalyst science as a basis for future developments in catalyst technology, and to provide an adequate supply of scientists educated in the principles underlying catalysis, the panel recommends, in order of importance, the following:

1. A materials-focused approach is needed to complement the existing strong efforts on understanding and elucidating catalytic phenomena. More emphasis should be placed on investigation of the optimized design and synthesis of new catalytic materials, in addition to the study of existing ones. It must be kept in mind that a new material deserves consideration as a potential catalytic material only after its successful use as a catalyst, or as a component of such.

2. Further advancement should be made in the characterization of catalysts and the elucidation of catalytic processes, particularly under reaction conditions; existing studies of structure-function relationships should be continued and expanded to focus on catalysts relevant to applications with major potential.

3. Academic researchers should develop cooperative, interdisciplinary projects, or instrumental facilities, in which researchers from a range of disciplines work on various aspects of a common goal, as exemplified by programs carried out in NSF-supported Science and Technology Centers.

4. Academic researchers should be encouraged to work collaboratively on projects with industry that are aimed at enabling the development of catalyst technology through the application of basic knowledge of catalysts and catalytic phenomena.

5. Academic institutions should ease their patent policies with respect to ownership and royalties, to facilitate greater industrial support of research.

For National Laboratories

The national laboratories have been highly effective in developing novel instrumentation for catalyst characterization, operating large-scale user facilities (i.e., synchrotron radiation sources, pulsed neutron sources, and atomic resolution microscopes), and applying the most advanced experimental and theoretical techniques to study structure-function relationships critical for understanding catalysis at the molecular level. Given the wealth of resources at these laboratories, major opportunities exist for advancing catalyst science and technology through research, including work carried out in collaboration with industry and academic scientists. The panel encourages the national laboratories to

1. undertake collaborative research with industry focused on developing fundamental understanding of the structure-property relationships of industrially relevant catalysts and catalytic processes, and on using such understanding for the design of important new catalysts;

2. continue the development of novel instrumentation for in situ studies of catalysts and catalytic phenomena;

3. place greater emphasis on the systematic synthesis of new classes of materials of potential interest as catalysts; and

4. investigate novel catalytic approaches to the production of energy, the selective synthesis of commodity and fine chemicals, and the protection of the environment.

For the Federal Government

The principal sources of support for university and national laboratory research on catalysis are the Department of Energy (DOE) and the National Science Foundation (NSF). As noted in Chapter 4, constant-dollar funding from these agencies, together with inflation and rising overhead costs, has caused a decrease in the number of young scientists being trained in the field of catalysis. The panel also observes that with the decline in emphasis on alternative fuels, research in catalysis has become increasingly diversified and less aligned along national interests. To offset these trends, the panel recommends that federal agencies

1. establish mechanisms for reviewing their programs related to catalysis, to ensure that they are balanced and responsive to the needs of the nation and to the opportunities for accelerating progress;

2. encourage industry to assist the funding agencies in identifying important fundamental problems that must be solved to facilitate the translation of new discoveries into viable products and processes; assessment of the fundamental research needs of industry should be communicated to all members of the catalysis community; and

3. increase the level of federal funding in support of catalysis research by at least a factor of two (after correction for inflation) over the next five years. This recommendation is consistent with the Bush administration's proposal to double the NSF budget over the next five years and with a recent statement by Frank Press, president of the National Academy of Sciences, that doubling the research budgets of all federal agencies

should be a goal for the 1990s. Recognizing the need for federal agencies to maintain flexibility and to encourage creative scientists who propose to explore new directions and ideas, the panel recommends that priority be given to the following five areas:

• **Synthesis of new catalytic materials and understanding of the relationships between synthesis and catalyst activity, selectivity, and durability.**

• **Development of in situ methods for characterizing the composition and structure of catalysts, and structure-function relationships for catalysts and catalytic processes of existing, and potential, industrial interest.**

• **Development and application of theoretical methods for predicting the structure and stability of catalysts, as well as the energetics and dynamics of elementary processes occurring during catalysis, and use of this information for the design of novel catalytic cycles and catalytic materials and structures.**

• **Investigation of novel catalytic approaches for the production of chemicals and fuels in an environmentally benign fashion, the production of fuels from non-petroleum sources, the catalytic abatement of toxic emissions, and the selective synthesis of enantiomerically pure products.**

• **Provision of the instrumentation, computational resources, and infrastructure needed to ensure the cost-effectiveness of the entire research portfolio.**

This report is intended to identify the research opportunities and challenges for catalysis in the coming decades, to document the achievements and impacts of catalytic technologies—past and still to come, and to detail the resources needed to ensure the continued progress that will enable the United States to remain a world leader in the provision of new catalytic technologies. Chapter 1 provides an introduction to the science and technology of catalysts. Chapter 2 discusses the opportunities for developing new catalysts to meet the demands of the chemical and fuel industries, and the increasing role of catalytic technology in environmental protection. The intellectual challenges for advancing the frontiers of catalytic science are outlined in Chapter 3. The human and institutional resources available in the United States for carrying out research on catalysis are summarized in Chapter 4. The panel's findings and recommendations for industry, academe, the national laboratories, and the federal government are presented in fuller detail in Chapter 5.

1

Introduction

The following are some benchmark discoveries made over the years in the science and technology of catalysis.

• 100 years ago: Paul Sabatier (Nobel Prize 1912) at the University of Toulouse started work on his method of hydrogenating organic molecules in the presence of metallic powders.

• 70 years ago: Irving Langmuir (Nobel Prize 1932) at General Electric laid down the scientific foundations for the oxidation of carbon monoxide on palladium.

• 50 years ago: Vladimir Ipatieff and Herman Pines at UOP developed a process to make high-octane gasoline that was shipped just in time to secure the victory of the Royal Air Force in the Battle of Britain.

• 30 years ago: Karl Ziegler and Giulio Natta (Nobel Prize 1963) invented processes to make new plastic and fiber materials.

• 17 years ago: W. S. Knowles at Monsanto Company obtained a patent for a better way to make the drug L-Dopa to treat Parkinson's disease.

• 16 years ago: General Motors Corporation and Ford Motor Company introduced new devices in cars to clean automotive exhaust. These devices found worldwide acceptance.

• 10 years ago: Tennessee Eastman Corporation started a new process for converting coal into chemicals used for the production of photographic film.

• Yesterday: Procter and Gamble Company manufactured a new environmentally safe bleach mixed with laundry soap.

• Today: Thomas Cech (Nobel Prize 1989) at the University of Colorado received U.S. patent 4,987,071 to make ribozymes, a genetic material that might, one day, be used to deactivate deadly viruses.

• Tomorrow: A new, homogeneous catalyst to make methanol may be commercialized following preliminary work at Brookhaven National Laboratory.

The above examples deal with materials for health, clothing, consumer products, fuels, and protection of the environment, but all have a common feature: they rely on chemical or biochemical catalysts. What are catalysts? What is catalysis, this growing field of science and technology that holds the keys to better products and processes, and continues to have such a strong impact on our economy and quality of life?

WHAT ARE CATALYSTS?

The word "catalyst" is often used in everyday conversation: it is said, for instance, that a person is a catalyst, meaning a go-between who facilitates a transaction but withdraws when the transaction is ended. Similarly, a catalyst is, in principle, found intact at the end of a chemical reaction, ready to be engaged in the same reaction again and again. What the catalyst

---

THE BATTLE OF BRITAIN: CATALYSTS FOR VICTORY

Fifty years ago, between July 10 and October 31, 1940, Royal Air Force fighter pilots defeated the Luftwaffe in a heroic air battle over Britain. The British lost 915 planes versus 1733 for the Germans. The impact of the British victory was immortalized by Winston Churchill in the House of Commons when he said, "Never in the field of human conflict was so much owed by so many to so few."

In the *Chicago Tribune Magazine* of July 15, 1990, Herman Pines reminds us of the critical role played by 100-octane fuel that provided British planes with 50% faster bursts of acceleration than were available to them during the May 1940 French campaign fought with 87-octane fuel. With the same planes but new fuel, British pilots were able to outclimb and outmaneuver the enemy.

The new fuels that contributed to victory came just in time from the United States, as a result of discovery and development by Universal Oil Products (now UOP Inc.) of sulfuric acid-catalyzed gasoline alkylation. Vladimir Ipatieff, Herman Pines, and Herman S. Bloch played key roles in this work.

Since 1940, hydrofluoric acid has, in part, replaced sulfuric acid as the catalyst for gasoline alkylation. Today, in the battle for the environment, efforts are under way to replace hydrofluoric acid. Eventual success will be another achievement of researchers in catalysis.

does is to provide a path for the reaction to proceed swiftly and selectively to the desired products. Yet what about an operational definition of a catalyst?

A catalyst is a substance that transforms reactants into products, through an uninterrupted and repeated cycle of elementary steps, until the last step in the cycle regenerates the catalyst in its original form. Many types of materials can serve as catalysts. These include metals, compounds (e.g., metal oxides, sulfides, nitrides), organometallic complexes, and enzymes. Because not all portions of a catalyst participate in the transformation of reactants to products, those portions that do are referred to as active sites. Most industrial catalysts are used in the form of porous pellets, each of which contains typically 10^{18} catalytic sites.

The total amount of catalyst is small compared to the amount of reactants and products made during the life of the catalyst. The turnover frequency of the cycle is the quantity that defines the activity of a catalyst. Strictly speaking, the turnover frequency is the number of molecules of a given product made per catalytic site per unit time. In heterogeneous catalysis, the turnover frequency is typically of the order of one per second.

Who develops these catalysts? The development of catalysts is carried out by chemists and chemical engineers, often in large multidisciplinary teams that bring together expertise in the areas of physical, organic, and inorganic chemistry, as well as materials science and chemical reaction engineering. Such teams work on determining the optimal composition and physical structure of the catalyst, its activity and selectivity over the desired range of operating conditions, and its deactivation rate over time. Attention is also paid to developing methods for catalyst reactivation and recovery.

The generalities cited above can be illustrated by examples borrowed from the chemical, oil, and pharmaceutical industries, as well as environmental protection.

The first triumph of large-scale catalytic technology goes back to 1913 when the first industrial plant to synthesize ammonia from its constituents, nitrogen and hydrogen, was inaugurated in Germany. From the outset, and until the present, the catalyst in such plants has consisted essentially of iron. The mechanism of the reaction is now well understood. Small groups of iron atoms at the surface of the catalyst are capable of dissociating first a molecule of nitrogen and then a molecule of hydrogen, and finally of recombining the fragments to ultimately form a molecule of ammonia. The catalyst operates at high temperature to increase the speed of the catalytic cycle and at high pressure to increase the thermodynamic yield of ammonia. Under these severe conditions, the catalytic cycle turns over more than a billion times at each catalytic site before the catalyst has to be replaced. This high productivity of the catalyst explains its low cost: the catalyst results in products worth 2000 times its own value during its useful life.

The refining of petroleum to produce fuels for heating and transportation involves a large number of catalytic processes. One of these is the catalytic reforming of naphtha, a component derived from petroleum, used to produce high-octane gasoline. In modern catalytic reforming, many different catalytic reactions proceed on small particles made of platinum and a second metal such as rhenium or iridium. These bimetallic clusters are expensive but chemically robust. They can be reactivated after long-term use, thus making possible the use of precious metals to produce an affordable consumer commodity. The metallic clusters are so small that practically all metal atoms are exposed to the reactants and take part in the catalytic cycle. These metal clusters are supported within the pores of an acidic metal oxide that also takes part in the reforming process.

The next illustration of catalysis shows that industrial catalysts can be biomimetic, in the sense that they imitate the ability of enzymes to produce optically active molecules (i.e., molecules whose structures are such that the reflection of the molecule in a mirror does not superimpose on the original molecule). Many pharmaceuticals are known to be active in only one form, let us say the left-handed form. It is therefore critical to obtain the left-handed form with high purity. This is particularly important when the drug is toxic, even if only slightly so, and must be administered over many years. It is true of a molecule called L-Dopa used in the treatment of Parkinson's disease. Here, the right-handed molecule is inactive. In ordinary synthesis, both forms (right and left) are produced in equal amounts. Their separation is costly. Is it possible to produce only the left-handed form by means of a synthetic catalyst? The first success of an industrial synthesis of this kind was achieved at Monsanto, and a patent for the selective synthesis of L-Dopa was granted in 1974. The catalytic process used to make L-Dopa today may be regarded as an important achievement in industrial catalysis.

Finally, more recent developments in catalytic technology are targeted at the protection of the environment. The best-known example deals with catalytic converters that remove pollutants from the exhaust gases of automobiles. Catalytic converters for automobiles were first installed in the United States in the fall of 1974. These devices were subsequently introduced in Japan and are currently spreading through Europe. The most advanced catalyst now contains three metals of the platinum group and controls the emissions of carbon monoxide, nitrogen oxides, and unburned hydrocarbon molecules by use of a complex network of catalytic reactions. This application has contributed more than any other to public awareness of catalysis and of its many applications for the benefit of mankind.

AN IMMOBILIZED ENZYME AS AN INDUSTRIAL CATALYST

While the oil crisis of the 1970s was front-page news, the soft drink industry was experiencing a less-heralded shock of its own. High sugar cane prices sparked a scramble to other sweeteners—not artificial sweeteners, mind you, but other forms of sugar. Table sugar—sucrose—is just one member of a family of several dozen closely related natural sugars. Other members include fructose, found in fruits; glucose, found in honey and grapes; lactose, found in milk; and maltose, found in malted grain. Fructose became the sweetener of choice, thanks to the adaptation of a catalyst to its large-scale industrial production. The catalyst, glucose isomerase, was derived from *Streptomyces*, a common, soil-dwelling bacterium best known as the source of many antibiotics, including streptomycin.

Glucose isomerase is an enzyme—one of nature's own catalysts. All living things use enzymes, each one tailored to carry out one of the multitude of reactions essential for life itself. An enzyme is essentially a protein molecule, although it may have other atoms or molecules attached that help it do its job. A protein molecule is a long chain made up of hundreds or thousands of smaller units called amino acids assembled in a very specific order. When dissolved in water, this chain naturally kinks and knots up. The sequence of amino acids making up the protein determines the shape that the protein knots itself into, and it is this shape that allows the protein to catalyze its reaction. The molecules that participate in the reaction fit into a crevice in the protein, like a key in its lock. Once inside the crevice, called the "active site," the molecules are held in just the right relative orientation for the reaction between them to proceed.

Adapting an enzyme to a continuous-flow industrial process requires that the soluble protein be immobilized somehow. (If the protein were left in its soluble form, it would be well-nigh impossible to separate it from the process stream, and it would all wash away in the flow.) To keep doing its job, the immobilized enzyme must retain its dissolved shape, yet it must also be firmly anchored to its solid support. In addition, the catalyst-support combination must be stable at the processing temperature and strong enough not to break up under processing conditions. Resin and polymer supports were tried first, because these molecules are chemically very similar to enzymes, which makes it easy to attach enzymes to them. Unfortunately, the resin and polymer beads were crushed into a gummy mass under the processing conditions and clogged the works.

One way around the problem is to attach the enzyme molecules to a ceramic material. Tiny ceramic particles have a high surface area, allowing a lot of catalyst to be attached to them and increasing the reaction's efficiency. Ceramics are also incompressible, and so the system can be run at high pressure without crushing the catalyst or clogging the works. Because proteins do not stick naturally to ceramics, an intermediary is needed. The ceramic particles are coated with a special polymer that adheres well to both the ceramic and the enzyme while allowing the enzyme to retain its shape.

The enzyme does tend to decompose slowly under process conditions, but the extra stiffness imparted to it by the ceramic backbone makes it more stable than the natural enzyme, so that each batch of ceramic-supported enzyme lasts longer. The fructose production process has proved to be remarkably efficient. One pound of catalyst-coated ceramic will produce an average of 14¼ tons, and sometimes as much as 18 tons, of fructose (measured as a dry solid) before the enzyme loses its activity.

The process converts a watery, honey-colored syrup containing 95% glucose—a by-product of the wet-milling process used to make starch from corn—into a 42-45% fructose syrup—the "corn sweeteners" on a soft drink ingredient label. (Both Coca-Cola and Pepsi-Cola allow their bottlers to replace up to 100% of the sugar in their soft drinks with corn sweeteners.) Although glucose, fructose, and sucrose are all sugars, they are not equally sweet. Glucose is not picked up by the taste buds as quickly as fructose or sucrose, nor does its sweetness linger as long on the palate. If sucrose scores 100 on a sweetness scale, fructose rates a supersweet 173 and glucose an unsatisfying 74—the main reason glucose itself is not sold as a sweetener.

A complicated separation process keeps recycling unreacted glucose back through the system, while drawing off fructose as it forms. Pure fructose comes out as a 90% solution, which is diluted to 55%—equivalent in sweetness to pure sucrose—before the syrup is sold. Thus fructose is as convenient to use as sucrose—the bottler does not have to install any extra tanks or plumbing to dilute the fructose, or alter the recipes to allow for its greater sweetness. Fructose has the added advantage of being safe for diabetics.

THE FUNCTION OF RESEARCH

Research plays a vital role in advancing the frontiers of scientific understanding of catalysis and in assisting the development of catalysts for industrial application. Because of the complexity of catalysts and catalytic phenomena, information must be drawn from a large number of supporting disciplines, in particular,

- organometallic chemistry,
- surface science,
- solid-state chemistry and materials science,

- biochemistry and biomimetic chemistry,
- chemical reaction engineering, and
- chemical kinetics and dynamics.

Although research in catalysis is still dominated largely by experimental studies, theoretical efforts are becoming increasingly important. Theory provides a framework for understanding the relationships among catalyst composition, structure, and performance. The advent of supercomputers has made it possible to model a still larger body of catalytic phenomena and even, in some cases, to predict catalyst properties a priori. The availability of high-resolution computer graphics has proved particularly useful in visualizing the results of complex calculations and understanding the spatial relationships between catalysts and reactants on a molecular scale.

The industrial development of catalysts is an expensive and labor-intensive activity because catalysts currently cannot be designed from first principles. Rather, they must be developed via a sequence of steps involving formulation, testing, and analysis. An important aim of research in catalysis is to accelerate this process by providing critically needed knowledge and techniques. Another important function of research is to provide a reservoir of new information and materials that may contribute to the identification of new catalytic materials or processes. Thus, not only does research provide the tools and knowledge needed for direct facilitation of catalyst development, but it also increases opportunities for the discovery of new materials and new techniques.

One example suffices to illustrate the impact of research on catalytic science and technology. Because catalysis is a kinetic phenomenon based on the turning over of the catalytic cycle, the example deals with the prediction of overall kinetics for a catalyzed reaction based on a knowledge of elementary steps in the cycle. This information cannot be obtained theoretically at present, but it can be determined from experimental investigations. In the case of solid catalysts, some of these measurements are carried out on large single crystals, exposing one defined facet about 1 cm in size. These facets are nearly perfect in structure and are extremely pure. The chemistry of elementary processes occurring on such surfaces can be studied in great detail, to determine not only the rate of the process but also what intermediate species are formed. By studying different crystal facets, the effects of catalyst surface structure on reaction dynamics can be established. Information on the rates of elementary reactions can be assembled to describe the kinetics of a multistep process. This approach has been used to understand ammonia synthesis, over iron, and to establish which facet of iron is most effective in promoting this reaction. Such knowledge can be used to guide the preparation of industrial catalysts so as to expose the desired facets of iron preferentially. Recent studies have demonstrated that the best commercially available ammonia synthesis catalysts operate at a rate that is almost equal to that observed on the preferred facet of iron.

SUMMARY AND PERSPECTIVE

In summary, catalysts play a vital role in providing society with fuels, commodity and fine chemicals, pharmaceuticals, and means for protecting the environment. To be useful, a good catalyst must have a high turnover frequency (activity), produce the right kind of product (selectivity), and have a long life (durability), all at an acceptable cost. Research in the field of catalysis provides the tools and understanding required to facilitate and accelerate the development of improved catalysts and to open opportunities for the discovery of new catalytic processes.

The aim of this report is to identify the research opportunities and challenges for catalysis in the coming decades and to detail the resources necessary to ensure steady progress. Chapter 2 discusses opportunities for developing new catalysts to meet the demands of the chemical and fuel industries, and the increasing role of catalysis in environmental protection. The intellectual challenges for advancing the frontiers of catalytic science are outlined in Chapter 3. The human and institutional resources available in the United States for carrying out research on catalysis are summarized in Chapter 4. The findings and recommendations of the panel for industry, academe, the national laboratories, and the federal government are presented in Chapter 5.

2

New Opportunities in Catalytic Technology

SOCIETAL IMPACT OF CATALYTIC TECHNOLOGY

The chemical industry is one of the largest of all U.S. industries, with sales in 1990 of $292 billion and employment of 1.1 million.[1] It is one of the nations's few industries that produces a favorable trade balance; the United States now exports chemical products amounting to almost twice the value of those that it imports (exports of roughly $37 billion compared to imports valued at about $21 billion).[2] Between 1930 and the early 1980s, 63 major products and 34 major process innovations were introduced by the chemical industry. More than 60% of these products and 90% of these processes were based on catalysis. Catalysis also lies at the heart of the petroleum refining industry, which had sales in 1990 of $140 billion and employed 0.75 million workers.[3] The very high degree of automation in the chemical and petroleum industries contributes to their lower labor intensity and to the fact that labor costs are much less important than material costs in the manufacture of petroleum refinery products and most chemicals. Clearly then, *catalysis is critical to two of the largest industries in sales in the United States; catalysis is also a vital component of a number of the national critical technologies identified recently by the National Critical Technologies Panel.*[4]

---

[1]U.S. Department of Commerce, *U.S. Industrial Outlook 1991*, International Trade Administration, Washington, D.C., 1991.

[2]U.S. Department of Commerce, *U.S. Industrial Outlook 1991*.

[3]U.S. Department of Commerce, *U.S. Industrial Outlook 1991*.

[4]*Report of the National Critical Technologies Panel*, William D. Phillips, chair, Arlington, Va., March 1991.

CATALYTIC CRACKING: MAJOR IMPACT ON THE
U.S. BALANCE OF PAYMENTS

During the refining of petroleum, large hydrocarbon molecules are broken down into smaller ones in a process known as cracking. The amount of gasoline that can ultimately be produced from a barrel of oil depends on how efficiently cracking is performed. If carried out incorrectly, cracking can lead to the formation of gases such as methane and ethane, and high-molecular-weight components called residua, which cannot be used to make gasoline or other transportation fuels. Today, through the use of highly optimized catalysts, more than 70% of the cracked products end up as transportation fuels.

The story began in 1936 when acid-washed natural clays were first employed as catalysts. Subsequent research revealed that higher cracking efficiencies could be achieved by using amorphous silica-alumina. In the late 1950s and early 1960s, significantly greater efficiencies were found to be possible by using cracking catalysts based on zeolites. These materials are crystalline solids containing pores and cavities of molecular dimensions. The interior surfaces of the zeolite contain highly acidic centers that serve as the active sites for cracking petroleum (Figure 2.1). Mobil Corp. introduced the first zeolite-containing cracking catalyst in the early 1960s, and today virtually every refinery in the world uses catalysts containing Y-zeolite and, in some instances, ZSM-5.

The use of zeolite catalysts has greatly benefited the U.S. balance of payments, because the improved efficiency in cracking has permitted a savings of more than 400 million barrels of oil per year, or more than $8 billion a year at $20 a barrel, and the story continues with new catalysts showing promise of even greater selectivity. For example, a mere 1% shift in product selectivity to gasoline allows a reduction in oil imports by more than 22 million barrels of crude per year or more than $400 million in the U.S. balance of payments.

Looking into the future, one can see many exciting challenges and opportunities for developing totally new catalytic technologies and for further improving existing ones. Increasing public concern with the effects of chemicals and industrial emissions on the environment calls for the discovery and development of processes that eliminate, or at least minimize, the use and release of hazardous materials. Concern with the environment and the supply of raw materials is also focusing attention on the opportunities for recycling. Of particular interest for the chemical industry is the prospect of producing polymers that are readily recyclable. Although the world supply of petroleum is adequate for current demand, there is a need to continue the search for technologies that will permit the conversion of

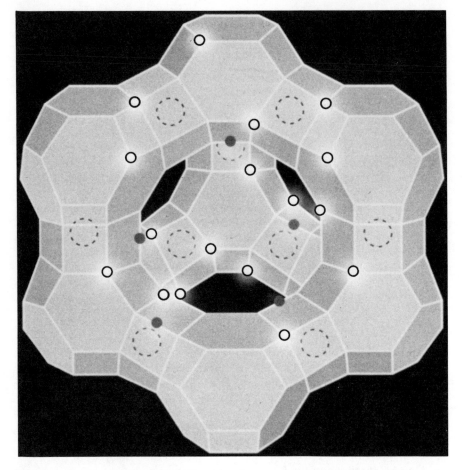

Figure 2.1   A representation of the molecular structure of HY zeolite. Solid circles represent the Brønsted acid sites responsible for cracking of petroleum. Open circles represent (Al O_4)$^-$ sites, and dashed circles represent Me$^+$ sites. (Figure courtesy of Union Carbide Corporation.)

methane, shale, and coal into liquid fuels at an acceptable cost. Also, to maintain economic competitiveness, it will be necessary to shift to lower-cost feedstocks for the production of commodity and fine chemicals. Taken together, these forces provide a strong incentive for increasing research efforts aimed at the discovery of novel catalysts and catalytic processes.

The markets for industrial catalysts are usually broken down into three sectors: chemicals, fuels, and environmental protection. The same classification is used in the sections that follow to discuss new opportunities in catalytic technology.

PRODUCTION OF CHEMICALS

Existing Products

In considering new routes to existing products, emphasis here is placed on major advances rather than on incremental improvements, even though the latter are often quite valuable and justified. With the increasing maturity of catalytic technology for most large-volume commodity chemicals, major advances in the future will require technical discontinuities. These discontinuities, as opposed to improvements in existing technologies, offer the real opportunities for catalysis to have an impact on the economy.

One can recognize and identify limits in the current technology for almost all major products made via catalytic processes. Furthermore, in most cases, at least one potential pathway to a major advance can be visualized. Each such advance constitutes a latent opportunity to shift to a lower-cost feedstock or to a simpler, less-capital-intensive route.

Lower-Cost Feedstocks

For typical commodity chemical processes, feedstock constitutes about 60-70% of the total manufacturing cost. Thus, a great financial impact can result from moving to a lower-cost feedstock. For example, the Monsanto process for making acetic acid (CH_3COOH) via methanol (CH_3OH) carbonylation involved a feedstock change—a shift from ethylene (C_2H_4), used in the previously dominant Wacker process, to methanol (CH_3OH). Since its launching in 1970, the Monsanto process has captured most of the world's new capacity for making acetic acid. Feedstock price changes in recent years have further magnified the cost advantage of the methanol carbonylation route.

Wacker Process: $(C_2H_4) + (H_2O) \xrightarrow{\text{air}} (CH_3CHO) \xrightarrow{\text{air}} (CH_3COOH)$
Monsanto Process: $(CH_3OH) + (CO) \rightarrow (CH_3COOH)$

By far the strongest current thrust toward lower-cost feedstocks is the effort to substitute alkanes (ethane, propane, and butane) for the corresponding olefins and to convert methane to olefins or aromatics. The difference in price between these alkanes and their olefin counterparts can frequently be as much as 10¢ per pound, which is substantial. An excellent example is the production of maleic anhydride, a monomer for specialty plastics. Over the past 40 years, advances in catalytic technology have enabled the industry to switch from high-priced, toxic benzene to butenes and, more recently, to the lower-cost hydrocarbon butane. This latter development was possible only as a result of the discovery of the vanadyl phosphate catalyst by Chevron Research and Technology Company. A new process for producing maleic anhydride by using novel catalyst and reactor technologies is currently under development by Du Pont and is scheduled for commercialization in the mid-1990s.

An extensive worldwide effort is now under way to develop the catalytic oxidative coupling of methane to petrochemicals as well as liquid fuels. This effort encompasses oxidative coupling of methane to ethylene or aromatics, oxidative methylation of toluene to ethylbenzene and styrene, oxidative methylation of propylene to C_4 olefins, and dehydrogenative coupling of methane to aromatics. This area of methane conversion captured exceptional interest and attention all over the world in the mid 1980s.

Thermal Cracking: $(C_2H_6) \rightarrow (C_2H_4) + (H_2)$
Oxidative Coupling: $(2CH_4) + (O_2) \rightarrow (C_2H_4) + (2H_2O)$

The relative abundance of LPG (liquefied petroleum gas, containing mostly propane and butane) and the strong demand for aromatics have prompted British Petroleum (BP) to develop a process for the catalytic conversion of LPG to aromatics. The BP process employs a zeolite-based catalyst developed by BP in conjunction with UOP's continuous catalyst regeneration system. LPG is converted to a mixture of aromatics, 95% of which are benzene, toluene, and xylenes. The aromatics yield is 65%. In another alkane utilization project, BP Chemicals is developing a process for the direct one-step ammoxidation of propane to acrylonitrile. Key to the process is a proprietary catalyst. Now at the pilot-plant stage, the process is targeted for commercialization in the mid-1990s.

A new commercial development in catalytic alkane dehydrogenation relates to the production of isobutylene and of propylene. The isobutylene requirement is for the production of gasoline octane enhancers (i.e., methyl tertiary-butyl ether, or MTBE), and the propylene need is driven by changes in the feedstock used to produce ethylene, which have resulted in less by-product propylene production. Several companies have recently installed or are currently installing new plants for the production of isobutylene and propylene. In light of this remarkable development, there may also be opportunities for new catalysts that would be capable of promoting oxidative dehydrogenation of lower alkanes (i.e., ethane, propane, and isobutane) to their corresponding olefins. Direct functionalization of hydrocarbons remains a very significant approach (i.e., ethane to ethanol, propane to acrylonitrile, and butane to 1,4-butanediol).

Catalytic dehydrogenation of paraffins is also widely practiced commercially for the production of linear olefins in the C_{10}-C_{17} range, used in the manufacture of biodegradable detergent intermediates. Typically, olefins in the C_{10}-C_{14} range are used in the production of linear alkylbenzenes (LAB) whereas heavier olefins in the C_{14}-C_{17} range are used in the preparation of detergent alcohols via hydroformylation. The preparation of yet heavier olefins by catalytic dehydrogenation is also possible for specialized applications, including the manufacture of synthetic lubricating oils and oil additives. Worth noting at this point is the recent introduction of solid (heterogeneous) acid catalysts for the alkylation of benzene with heavy olefins in the production of LAB; this will allow the replacement of traditional catalysts, such as hydrogen fluoride (HF) or aluminum chloride ($AlCl_3$) used for this purpose and will thus avoid the operational hazards associated with the handling and processing of corrosive catalysts and ameliorate the environmental characteristics of this alkylation process.

C_1 chemistry (i.e., chemical processes based on carbon monoxide, carbon dioxide, or methanol as the starting material) now provides another interesting arena for feedstock-driven innovations in industrial catalysis. After the oil embargo of 1973, there was an extensive worldwide effort to pursue C_1 chemistry for the production of chemicals as well as fuels. This effort eventually subsided when it appeared that the cost of carbon from C_1 sources such as coal and natural gas could not really compete effectively with its cost from petroleum-based sources, even at the much inflated prices of the latter. However, it appears that some significant changes have occurred in the past decade (before Iraq invaded Kuwait) and that the opportunities for making chemicals via C_1 chemistry should be revisited. In particular, methanol should be considered as a feedstock. Figure 2.2 illustrates the historical and forecasted trends between 1955 and 1998 for the ratio of methanol to ethylene prices. A substantial downward trend in favor of methanol can be observed.

It has recently been reported that rhodium-based homogeneous catalysts promote the reductive carbonylation of methanol to acetaldehyde at selectivities approaching 90% and at much lower pressure than required for prior-art catalysts. With the addition of ruthenium as co-catalyst, it is possible to achieve in situ reduction of acetaldehyde to ethanol, thus providing a new catalyst system for the homologation of methanol to ethanol.

One great challenge for catalysis has been the possibility of producing ethylene glycol via the oxidative coupling of methanol rather than the standard process based on ethylene as feedstock. Significant progress has been made recently in the catalytic oxidative dimerization of dimethyl ether to dimethoxyethane. Dimethoxyethane, in turn, should be hydrolyzable to ethylene glycol. Unlike methane coupling, which requires a temperature in excess of 600° C, the oxidative coupling of dimethyl ether proceeds at about 200° C with a mixed magnesium-tin oxide catalyst. By use of the ether rather than methanol, protection against side reactions has been achieved. These results are an extremely interesting lead which, coupled with the favorable trends in methanol pricing, could pave the way to another major advance in industrial catalysis (i.e., the production of ethylene glycol from dimethyl ether). In addition to acetaldehyde, ethanol, and ethylene glycol, other large-volume chemicals currently made from ethylene or propylene may become attractive candidates for manufacture via C_1 chemistry.

New Catalytic Oxidation Processes

Of the different classes of catalytic reactions, hydrocarbon oxidation (i.e., reaction with oxygen) is the one that generally has the lowest selectivity. In addition to the desired partial oxidation products, significant quantities of carbon monoxide, carbon dioxide, and water are often obtained. This results in complex and costly separations that, in turn, lead to processes with unusually high capital intensity. The annual capital expenditure for oxidation processes, per annual pound of product, is usually several times that for nonoxidative catalytic processes.

Well-recognized examples of major opportunities include the one-step

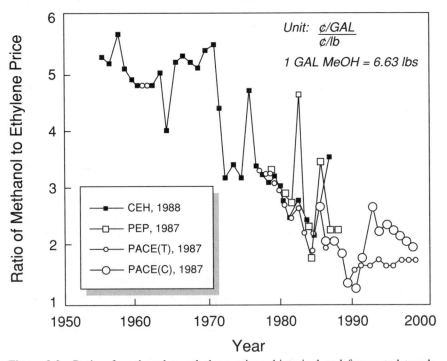

Figure 2.2   Ratio of methanol to ethylene prices, historical and forecasted trends, 1955 to 1998. (Reprinted, by permission, from J. Roth, 1991, p. 8 in *Catalytic Science and Technology*, Vol. 1, S. Yoshida, N. Takegawa, and T. Ono, eds., Kodansha Tokyo. Copyright © 1991 by Kodansha Ltd.)

oxidation of methane to methanol, of higher alkanes to alcohols, of propylene to propylene oxide, and of benzene to phenol. Recent reports of success in these areas are tantalizing and suggest that catalytic oxidation should be one of the most important and fruitful areas for innovation in industrial catalysis. We seem to be at the threshold of several discontinuous advances.

New Products

New developments in catalysis can and will be the enabling technology that gives rise to new products in many sectors of the chemical industry. The potential impact of catalysis on new products is illustrated in the areas of polymers, pharmaceuticals, and biologically derived products.

Polymers

The production of raw polymers (e.g., pellets) in the United States approached 60 billion pounds per year in 1990, which corresponds to a $30 billion business. In terms of fabrication into end-use articles (e.g., synthetic fibers, films, containers, structural parts), the polymer business is significantly larger. Catalysis contributes to both monomer and polymer synthesis for a major part of this industry.

Today, the United States has a clear advantage in polymer science. This position now yields a positive balance of trade but is undergoing significant competition from developments in Western Europe and Japan. To ensure a continued prominent position, rather substantial advances in catalyst technology for both monomers and polymers will be required.

Every polymer scientist involved with synthesis or structure-property studies has a "wish list" for new monomers or polymers that have not yet been able to be synthesized via clear-cut economic routes for commercial practice. Almost every family of polymeric materials can utilize advances in catalysis, in monomer production or in the polymerization process. Changes in material requirements, environmental issues, feedstock availability and economics, and worldwide competitive pressures make future catalytic advances extremely important.

The primary area of intense catalytic activity involves the synthesis of new or improved polyolefins. This industry evolved out of the original Ziegler-Natta catalyst discovery in the 1950s, leading to tens of billions of pounds per year of polyolefins worldwide. New catalyst breakthroughs could lead to new markets of significant volume—including diversified products such as syndiotactic polypropylene, true thermoplastic elastomers (e.g., propylene-elastomeric polypropylene resin (EPR)-propylene block co-

A STRONG POINT FOR THE FUTURE

Skyscrapers and bridges make our cities what they are. Airplanes, boats, and automobiles carry us anywhere in the world. Appliances fill our homes. Steel is the prime structural material in all these things. Steel has become as fundamental to our world as stone was to the caveman's. Going from stone tools to iron ones was a technological shift that must have taken considerable adjustment.

Just as our ancestors made the leap from stone to metals, so are we moving from metals to polymers (plastics). Although our heightened expectations of technology have made this transition much less traumatic, it is every bit as far-reaching. In the past 50 years, polymers have become ubiquitous; we wear them (polyester permanent-press fabrics), we walk on them (polypropylene shoe soles), we ride on them (polyisoprene tires), we sit on them (polyurethane sofa cushion stuffing, plastic furniture), we package our food in them (polyethylene plastic bags), and we even feed them to our computers (poly(vinyl acetate) floppy disks). Yet, who would have thought that polymers—which we tend to think of as soft, stretchy, and pliable—would ever challenge steel as a structural material?

When certain polymers are spun into fibers, the resulting materials are truly amazing. All polymers are long, chainlike molecules made up of many smaller molecules linked together. In flexible polymer fibers, such as polyester, the chains are partially relaxed—they are aligned along the direction of the fiber, but also loosely intertwined like spaghetti dangling from a fork. The chains in these new rigid polymers are fully extended and lie parallel to each other, like a fistful of uncooked spaghetti. These fibers are unexpectedly strong for their weight. One such fiber, poly(*para*-phenyleneterephthalamide), was found to have a tensile strength higher than that of a steel fiber of the same dimensions, yet it weighed one-fifth as much. The commercial development of this product, Kevlar, by Du Pont, was a long and very expensive process. More than twelve capital-intensive steps are required to convert the basic aromatic feedstocks into a strong polymer. A combination of acid catalysts, hydrogenation catalysts, and oxidation catalysts makes up the eight catalytic steps in this complex process, which required extensive engineering beyond the initial laboratory work to develop the chemistry into a viable commercial process.

Kevlar, this new kid on the structural materials block, has already made a dent in the automotive industry, replacing the steel belts in radial tires. Its strength, impact resistance, and light weight mean that cars and planes made largely of Kevlar may be in our future. Also, in terms of impact resistance, bullet-resistant vests worn by soldiers and policemen today are made of Kevlar (Figure 2.3). A thin vest of Kevlar can be worn comfortably inside one's shirt, yet it can stop slugs as effectively as a steel plate. This light, yet strong, material is only beginning to be used for the myriad applications that await it.

Figure 2.3 Photo of helmet and body armor made of Kevlar. (Figure courtesy of E. I. Du Pont de Nemours and Co., Inc.)

polymers), and incorporation of polar monomers into various polyolefin classes. For example, if acrylates, vinyl esters, acrylonitrile, and the like could be incorporated into the present low-pressure polyolefin synthesis, the result would be a new family of olefin-based polymers that are likely to have major commercial significance. Of course, improvements in the present catalytic systems will have a pronounced effect on the polyolefins that are commercially available. Improvements in molecular weight distribution control (e.g., narrow molecular weight distribution), the ability to synthesize EPRs in gas-phase reactors, and the control of catalyst decay (e.g., improved efficiency) are advances that will surely occur.

Novel metathesis catalysts for the synthesis of cyclic olefins have resulted in a host of new polymeric structures. Several of these have reached commercial status (e.g., *trans*-poly(octenamer)). The few catalysts that are effective are often expensive and require unattractive precautions during industrial scale-up. Further advances could make this a promising area for industrial exploitation.

In the field of functionalized monomers, the trend will likely revolve around new or existing monomers via biomass, coal, or C_1 conversion, as the carbon source availability changes together with economics. Improved catalysts for creating monomers from agricultural commodities or waste materials offer increasingly important opportunities. This field could benefit from the genetic engineering of specific enzyme "catalysts." Oxidative coupling of methanol to ethylene glycol and ethanol to 1,4-butanediol could open new routes to these important monomers. A new process (in progress) involving the one-step ammoxidation of propane to acrylonitrile could change the commercial position. Other possibilities—for example, utilizing shape-selective catalysts such as zeolites—could yield lower-cost routes to 4,4'-diphenol via phenol coupling. Such a molecule is of interest for liquid crystalline and engineering polymers. Improved non-phosgene routes to diisocyanates are desired. Functionalized oligomers such as hydroxyl (OH)-, amino (NH₂)-, and carboxyl (COOH)-terminated polyolefins could yield important blocks for step-growth polymers (e.g., urethanes, polyesters). Functionalized fluoroolefin oligomers for inclusion in step-growth polymers could likewise offer a new variety of polymers. A non-chlorine route to siloxane polymer precursors is also desired.

In the field of polymers, a number of prior failures have been well documented in the literature, primarily resulting from the unavailability of appropriate catalysts. The reaction of acetaldehyde to yield polyvinyl alcohol is one such example. Because phosgene is highly toxic, a non-phosgene route to polycarbonate is desirable. The polymerization of phenol to a highly linear unsubstituted poly(1,4-phenylene oxide) of high molecular weight would be of interest, as would the polymerization of polyphenylene by benzene coupling. There are many other examples of novel polymers waiting to be polymerized from available monomers.

Cationic polymerization continues to be an area of increasing research made possible by improved Lewis or Brønsted acid catalysis. Continued improvement is desired to yield higher molecular weights and industrially acceptable process conditions. Extension to other monomers (e.g., vinyl acetate) would be of future interest. Anionic polymerization is of interest primarily for unique block copolymers. Extension to additional monomers and the resultant block structures deserves more attention.

Pharmaceuticals

There can be no question that the 1990s will be the "decade of chirality." Many of the opportunities and challenges in this explosively evolving field stem from the pharmaceutical area and the growing recognition that the "wrong" enantiomer of a racemic drug represents a "medical pollutant" whose toxic side effects can far outweigh the therapeutic value of the pharmaceutically active enantiomer. The classic example in this area is that of thalidomide (Figure 2.4). The R-isomer of thalidomide is an effective sedative; tragically, the drug was sold as the racemate, and it was subsequently discovered that the S-isomer is a powerful teratogen. More recently, Eli Lilly was forced to withdraw its Oraflex anti-inflammatory because of liver damage caused by the "inactive" R-isomer. Although recent regulatory changes by the Food and Drug Administration stopped short of requiring that all drugs be sold as a single enantiomer, there is an obvious trend in this direction by drug companies.

Among the available strategies for the manufacture of optically pure substances, asymmetric catalysis provides powerful and unique advantages. Perhaps the foremost is the "multiplication of chirality"—the stereoselective production of a large quantity of chiral product by using a catalytic amount of a chiral source. Unlike fermentation, asymmetric catalysis is characterized by generality: processes are not limited to "biological"-type substrates, and the R- and S-isomers are made with equal ease. Asymmetric catalysis also circumvents the disposal of large amounts of spent nutrient media that are generated during fermentation. By comparison, optical resolution (i.e., diastereomeric crystallization) is extremely labor-intensive and necessarily produces 50% of the "wrong" isomer, which must be destroyed or racemized in a separate step.

Given the increasing importance of enantioselective synthesis, it is important that the United States place greater emphasis on this area. At present, Japan and the European Community are the leaders in basic research discoveries and applications.

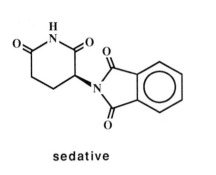
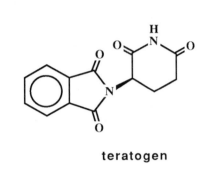

sedative   **teratogen**

Figure 2.4   Enantiomeric pairs of thalidomide.

Biologically Derived Products

The rapidly growing field of biotechnology brings with it opportunities in the field of enzyme-catalyzed reactions. The role of genetically engineered microorganisms in synthesizing rare and valuable peptides used in human therapeutics is now well established. The same techniques of molecular biology can also be used to enhance the properties of enzymes as catalysts for industrial processes that are very similar to classic catalytic technology.

This approach can potentially revolutionize the applications of biological systems in catalysis. Enzymes and other biological systems work well in dilute aqueous solution at moderate temperature, pressure, and pH. The reactions catalyzed by these systems are typically environmentally friendly in that few by-products or waste products are generated. The catalysts and the materials that they synthesize are, as a rule, biodegradable and therefore do not persist in the environment. The reactions are typically selective with extremely high yields, and enzymes can be used to catalyze a whole sequence of reactions in a single reactor, resulting in vastly improved overall yields with high positional specificity and 100% chiral synthesis in most cases. The improved use of enzyme catalyst technology with whole-cell catalysis, reactions catalyzed by single enzymes, and mixed enzymatic and chemical syntheses are all important for the development of new catalyst technology.

Whole cells of various microorganisms are being used more frequently in the catalytic synthesis of complex molecules from simple starting materials. The use of whole microbial cells as biosynthetic catalysts takes advantage of one of the unique properties of enzymes: they were designed by nature to function together in complex synthetic or degradative pathways. Because of this property, whole cells and microorganisms can be used as catalytic entities that carry out multiple reactions for the complete synthesis of complex chiral molecules. A patent was recently issued for a genetically engineered *Escherichia coli* that synthesizes the molecule D-biotin directly from glucose. Biotin has three chiral centers, and the current chemical synthesis requires 13-14 steps with low yields. Similarly, researchers are constructing a microorganism that directly catalyzes the synthesis of a vitamin C precursor from glucose. Combining genes from various organisms results in a process that uses a microbially synthesized intermediate with a final chemical conversion to vitamin C. Whole cells of microorganisms are also used in the synthesis of antibiotics from carbohydrate starting materials, and whole cells are used in the biocatalysis of certain steroids. A number of specialty chemicals with complex synthetic schemes can be produced most efficiently by intact microorganisms utilizing a series of enzyme-catalyzed reactions designed by nature to work together.

The biotechnology field also has a growing number of examples of reactions of industrial significance catalyzed by isolated enzymes. The conversion of cornstarch into corn syrup by the enzymes alpha- and gluco-amylase and glucose isomerase is a large industrial process, generating corn sweetener for soft drinks and other uses. The enzymatic conversion of acrylonitrile to acrylamide has recently been commercialized in Japan. Japanese companies and researchers have been very diligent in developing enzymatic processes for the synthesis of fine chemicals. Enzyme-catalyzed reactions are used by the Japanese for the synthesis of monosodium glutamate, L-tryptophan, and phenylalanine.

The stereospecificity of enzyme-catalyzed reactions has been used to advantage in polymer synthesis as well. Workers at ICI have developed a combined enzymatic and chemical process for the synthesis of polyphenylene from benzene. Benzene is oxidized to a *cis*-dihydrodiol by an enzyme-catalyzed oxygenation. The diol is derivatized, polymerized, and rearomatized to give polyphenylene in a reaction that cannot be carried out by classic chemical methods because of solubility problems. This new route to polyphenylene is an excellent example of combined enzyme and classic chemical synthesis to make a product that is otherwise too expensive for practical use. Other biological polymers are also finding their way into the catalyst field in various applications. Microorganisms are used to synthesize materials such as poly(beta-hydroxybutyrate), a biodegradable plastic, and researchers are exploring a series of synthetic silklike materials that may have uses in high-tensile-strength applications.

PRODUCTION OF FUELS

Existing Fuels

Although the current understanding of how the individual components of gasoline affect various environmental issues is very limited, several components in gasoline are now considered harmful to the environment if released to the atmosphere in high concentration as either spills, vaporization losses, or the result of incomplete combustion. They include aromatics, notably benzene, a carcinogenic reagent; high-vapor-pressure hydrocarbons such as butane; reactive hydrocarbons such as olefins; and sulfur compounds, which could promote the formation of smog and acids. The petroleum industry is responding to these concerns by directing its research toward the reformulation of gasolines. As discussed below, gasoline reformulation will require a number of advances in catalytic technology.

Advanced Fluidized Catalytic Cracking Catalysts for the Production of Environmentally Acceptable Gasoline

Innovations in catalytic cracking catalysts over the past 30 years have improved the conversion of the heavier components of crude oil to gasoline and diesel oil, allowing a reduction in crude imports of more than 400 million barrels a year. In the future, new types of cracking catalysts will be required to produce motor fuels that are environmentally more acceptable. To reduce the aromatics content of gasoline while at the same time providing higher-octane paraffinic components will require that fluidized catalytic cracking (FCC) catalysts produce more olefins. The highly reactive olefins can be isomerized, oligomerized, or alkylated with paraffins as well as reacted with methanol to produce a variety of high-octane ethers. For example, the zeolite ZSM-5 has been shown to be active in producing olefins when mixed with conventional faujasite-type cracking catalysts. ZSM-5 and other molecular sieves have also been proved active in isomerizing and oligomerizing olefins to a variety of liquid fuels.

Cracking catalysts will also have to become more rugged to endure the higher temperatures required to produce more olefins for subsequent processing into high-octane gasoline components. The catalytic cracking process has been the workhorse of modern refineries for 30 years and, with improved catalysts, will continue to be the main process for converting the heavier end of crude oil into more environmentally acceptable components for gasoline and diesel fuels.

Oxygenates for Octane Boosting

The need to remove aromatics from gasoline has created a need for organic oxygenates as replacement octane enhancers. Today the two predominant oxygenates used as octane enhancers are methyl tertiary-butyl ether (MTBE) and ethanol (EtOH). However, other oxygenates (e.g., alcohols, ethers, acetates, and carbonates) are known octane boosters. Some of these compounds have been evaluated only partially for their performance characteristics and may represent major growth opportunities in oxygenated fuels. Currently, approximately 150,000 barrels per day of oxygenated compounds (MTBE, EtOH) is added to gasoline in the United States. By the year 2000 it is projected that 750,000 barrels per day (the equivalent of roughly 10% of current U.S. production of petroleum) of oxygenates will be required for the gasoline pool. In the United States, the 1990 MTBE production capacity was 117,200 barrels per day. By 1993 it is targeted at 256,400 barrels per day. Iso-olefins (e.g., isobutylene) are now produced as a by-product of fluidized catalytic crackers in petroleum refineries. These quantities will support production of ether at a rate of only 200,000-300,000 barrels per day.

The process technology leading to the majority of oxygenates now in use has been made very efficient through catalyst modifications and engineering design. However, new processes leading to the same oxygenates may have cost advantages if different building blocks and feedstock sources (crude oil versus coal versus natural gas) are utilized.

The precursors of established octane enhancers—alcohols and ethers—rely heavily on natural gas and crude oil as their feedstock supply. With the United States importing approximately 55% of its crude oil, the price of oxygenates will track that of crude oil, which is currently approximately $20 per barrel. Industry estimates suggest that when crude oil prices pass approximately $30 per barrel, coal gasification to syngas and the conversion of syngas to hydrocarbons and oxygenates become cost competitive. Therefore, indirect liquefaction of coal offers promise for oxygenated fuels for the transportation industry.

New Fuels—Methanol Dissociation to Carbon Monoxide and Hydrogen

Methanol dissociation to carbon monoxide (CO) and hydrogen (H_2) on board a vehicle would provide a fuel that is even cleaner and more fuel efficient than undissociated methanol. The heat required for endothermic dissociation of methanol can be supplied readily by engine exhaust gas. This recovers waste heat and increases the heating value of the fuel. Fuel efficiency is further enhanced because an internal combustion engine running on dissociated methanol can be operated with excess air (i.e., lean combustion). Lean and complete combustion would ensure low CO and hydrocarbon emissions. The problems associated with formaldehyde emission would be significantly reduced. Reduction of nitrogen oxide (NO_x) emissions by an order of magnitude has been demonstrated because of lower combustion temperatures.

Dissociated methanol can also be used as an efficient peaking gas turbine fuel utilizing the heat recovery from exhaust gas. The endothermic methanol dissociation reaction is also promising for cooling critical systems such as engine cooling for hypersonic jets when dissociated methanol is used as a fuel. Methanol can be produced on a large scale in a remote location for shipping. The methanol dissociation reaction can be modified to yield a wide range of H_2 to CO ratios. It would provide an economically competitive and convenient source of CO or H_2, especially on a small scale, and could be used in chemical plants, materials processing plants, and fuel cells on board a vehicle.

Because of these emerging applications, there is renewed interest in developing methanol dissociation technology. Insufficient catalyst activity at low temperature and catalyst deactivation have been reported. These are two major challenges for catalysis research for on-board applications. Generally, copper-based catalysts are used at 250-300° C. Zinc-chromium and precious-metal-based catalysts are utilized at 350° C or higher. Improvements in catalyst performance are being sought. For example, the activity of copper-based catalysts can be increased by adding appropriate promoters and improving pretreatment environments. Catalyst stability depends on reaction temperatures. Active catalysts would allow reaction at lower temperature and would improve catalyst stability. Methanol dissociation could operate at near atmospheric pressure for passenger cars. Extension to about 15-20 atm is needed for turbine applications and to more than 100 atm for diesel engines.

ENVIRONMENTAL PROTECTION

Public interest in protecting the environment has increased and expanded greatly. These public concerns manifest themselves in many different ways. The challenge is to preserve the benefits of modern technology without seriously contaminating the natural world.

Three strategies are available for reducing the impact of chemicals on the environment: waste minimization, emission abatement, and remediation. Waste minimization calls for the design and development of products and processes that are inherently low-polluting or nonpolluting. The abatement of emissions can often be achieved by trapping harmful effluents or converting them to harmless substances (e.g., conversion of nitric oxide to nitrogen). Where an environmental insult has occurred, effective means of remediation are needed to restore the environment to its "green" state. As shown by the examples presented below, catalysis can contribute to these three approaches.

CATALYSIS FOR ENERGY INDEPENDENCE

The long gas lines and soaring prices during the Arab oil embargo of 1973 and the turmoil in Iran in 1979 made the U.S. public painfully aware of the nation's growing dependence on foreign oil. As the crises dissipated, however, so did public concern. Few lessons appear to have been learned. Today, the United States imports more oil than it did a decade ago—a fact brought home to the public once again by Iraq's takeover of Kuwait in 1990.

A few people did take a hint from the oil shocks of the 1970s. One oil company has developed a process to convert methanol (methyl alcohol, also called wood alcohol) to gasoline that stands as the first significant advance in synthetic fuel technology since the German program during World War II. The process converts natural gas, coal, or any carbon-rich raw material to premium gasoline. The carbon source is first "gasified," converted to a mixture of carbon monoxide and hydrogen gas by heating in the presence of steam and a catalyst. Another catalytic procedure converts the carbon monoxide and hydrogen to methanol—a standard industrial process. Methanol is converted directly to 95-octane gasoline through the action of a synthetic catalyst called ZSM-5.

ZSM-5 is a molecular sieve, which is to say that its three-dimensional crystal structure has an open, lacy quality to it—more hole than crystal. These pores are approximately 5 angstroms (about one-hundred billionth of an inch) in diameter, big enough to accommodate a methanol molecule. ZSM-5 catalytically breaks down the methanol molecule, whose chemical formula is CH_3OH, to create a molecule of water and a hydrocarbon fragment (CH_2). These fragments assemble themselves into hydrocarbon molecules—the stuff that gasoline is made of. (Gasoline is actually a rich stew of assorted hydrocarbon molecules, ranging from 6 to 12 carbon atoms per molecule.) All zeolites are molecular sieves, and many of them also catalyze this reaction, but the other zeolites keep adding the hydrocarbon fragments on to the gasoline-sized molecules, turning them into heavier and heavier liquids, and eventually into a solid. The pores in ZSM-5, however, are just the right size to construct gasoline-sized molecules, but too small to let larger hydrocarbons form. (Some of the intermediate products formed en route to gasoline-sized hydrocarbons are important chemicals in their own right. These chemicals can be intercepted as they form, and diverted for other uses. These uses include making high-quality synthetic lubricating oils and diesel fuels.)

The process was commercialized in 1985, when a plant was built on New Zealand's North Island. This facility converts offshore natural gas, mostly from the extensive Maui Gas Field in the Tasman Sea, into 14,500 barrels of gasoline per day. This gas-to-gasoline plant is designed to supply one-third of New Zealand's total liquid fuel demand and is a key to increasing its energy self-sufficiency.

Alkylation Catalysts

The alkylation of paraffins with olefins is one of the major refinery operations. The process reacts a paraffin, usually isobutane, with olefins, generally propylene and butenes, to produce highly branched C_7 and C_8 paraffins, respectively. They constitute a premium high-octane gasoline component (95-98 research octane number for C_4 alkylate). The alkylation process is assuming increasing importance with increased olefin production from modern fluid catalytic cracking units and with recent emphasis on clean fuels of lower aromatics content.

Both of the currently used liquid acid catalysts, sulfuric acid and hydrogen fluoride, are very corrosive. Acid waste disposal in the sulfuric acid-catalyzed process is of increasing environmental concern, and liquid hydrogen fluoride is a potential health hazard. With increasing concerns and possible legislative action addressing environmental and safety issues, current alkylation processes may face critical scrutiny.

Exploratory studies have shown that new catalysts can be developed that are cleaner and safer than those presently used. However, it is also apparent that powerful acid catalysts are required for alkylation. It is a formidable challenge to produce a novel catalyst system that makes a new process economically feasible: high yield of alkylate, selectivity to produce high-octane gasoline, long life cycle, regenerability, and greatly reduced environmental and safety risks. This is a challenge that would benefit from broad-based fundamental studies of acid catalysis and from increased exploratory research. At the present time, no satisfactory solid alkylation catalyst exists.

Replacements for Chlorofluorocarbons

Chlorofluorocarbons (CFCs) are now believed to contribute to the seasonal ozone depletion over the Antarctic continent. However, because they are crucial to many aspects of modern society and have no available replacements, it is not practical to cease their production immediately. By 1988, total CFC consumption worldwide had grown to 2.5 billion pounds per year. The three major uses in the United States are as refrigerants (30%), foam blowing agents for polystyrene and polyurethane (28%), and industrial solvents and cleaning agents (19%).

Ironically, it is the high stability and inertness of CFCs, which make them so valuable, that has led to their downfall. Once released at ground level into the atmosphere, they rise slowly into the stratosphere where they are degraded by high-energy radiation from the sun, to release chlorine-containing free radicals that trigger a catalytic ozone depletion cycle. Following detailed though not definitive studies, agreement was reached on a major global environmental treaty, the Montreal Protocol, to phase out CFC production by the turn of the century. A race by all CFC producers then began to find suitable and environmentally acceptable substitutes. The strategy is to reduce its atmospheric lifetime by introducing hydrogen into the molecule so that it is removed from the atmosphere by reaction with hydroxyl radicals in the troposphere. The commercially viable synthesis of these new compounds is a major challenge for catalysis, because catalysts used for the production of CFCs lack the required selectivity and activity to be acceptable for the production of hydrogen-containing substitutes.

The projected costs for these molecules, hydrogenated chlorofluorocarbons (HCFCs), are approximately 2-5 times those of the CFCs they are replacing, because of the complexity of the new manufacturing processes. Although rapid progress is being made toward the production of HCFCs, the latter are not entirely satisfactory and may have to be phased out in turn. Consequently, major advances in catalytic science and technology will be required to develop more acceptable substitutes before the turn of the century.

Emission Abatement

Catalytic technology is playing an ever-increasing role in environmental protection. In 1989, for the first time, the U.S. market for emission control catalysts (largely for automotive emissions) exceeded the market for petroleum refining catalysts. However, the area of stationary emission control (e.g., from power plants) has been flagged as one that will experience very large (20% per year) growth in the years ahead. There is also a need to reduce emissions from many chemical production plants. Thus, novel catalysis will in many cases be the critical technology that enables us to retain most of the benefits created by the chemical and petroleum industries, but with improved preservation of the environment.

Catalysts for automotive emission control are now well developed in the United States and, in general, meet mandated standards for removal of hydrocarbons, CO, and NO_x. Recent Clean Air Act revisions will require significantly greater reductions in 1993 in the emissions of hydrocarbons, CO, and NO_x, than those now mandated. In addition, very stringent local automobile emission standards (e.g., in California and Vermont) will require up to a 10-fold reduction in emissions by the late 1990s. This will necessitate more active catalysts, new catalyst supports (e.g., metallic supports), and new reactor designs that enhance low-temperature performance. A first step in this direction is the electrically heated converter, which offers a severalfold reduction in emissions over currently available technology. Because of the rapidly escalating price of rhodium (which promotes

AUTO EXHAUST CATALYSTS

Automobile exhaust is a well-known pollution source. It contains products of incomplete combustion—carbon monoxide and assorted light hydrocarbons—as well as combustion by-products such as oxides of nitrogen and sulfur. Catalytic converters that transform incomplete combustion products to carbon dioxide and water, and oxides of nitrogen back to nitrogen and oxygen, are required equipment on all automobiles sold today in most industrialized countries. Manufacturing these catalytic converters is a worldwide business worth several hundred million dollars per year. The oxidizing converters that finish off the incomplete combustion products were first commercialized in the United States in the fall of 1974, and "three-way" converters that also handle nitrogen oxides became available, again in the United States, in the fall of 1980. Recent tests of a new catalytic converter design show that substantial further reductions in hydrocarbon and carbon monoxide emissions may well be possible to meet the stringent requirements of the 1991 Clean Air Act.

Hydrocarbon and carbon monoxide emissions are at their highest levels during the first 10 minutes after the engine has been started. The cold engine burns fuel less efficiently, passing more incomplete combustion products on to the cold catalytic converter, which is itself less efficient. Heating the catalyst would make it more efficient faster, and the new design does just that. The converter is made of catalyst-coated stainless steel, which can be heated by passing an electric current through it (Figure 2.5). Tests of a prototype under the federal test procedure specified for catalytic converters have already shown substantial emission reductions in comparison to conventional technology. The electrically heated catalytic converter is being developed to meet the ultralow emission regulations enacted by California in late 1990 to help highly polluted areas such as the Los Angeles Basin.

NO_x reduction), there is a need for the development of lower-cost NO_x reduction catalysts or the development of practical catalysts for NO_x decomposition. With cleaner gasolines, there may also be opportunities to develop lower-cost catalysts containing a smaller amount of the expensive noble metals. Finally, with the very high cost of rhodium, there is need for an improved technology for recovering it.

The control of power plant emissions is another major area of opportunity for catalysis. In particular, there is a need for removal of NO_x emissions either via selective catalytic reduction (SCR) or, if achievable, via NO_x dissociation. The abatement of NO_x from power plants is important in efforts to control acid rain and photochemical smog, the latter being linked

Figure 2.5 Electrically heated automobile catalytic converter. (Figure courtesy of W. R. Grace & Company.)

with harmful ozone production. SCR removes NO_x in flue gas by reacting oxides of nitrogen with ammonia to form nitrogen and water. SCR was first commercialized in Japan and is now being used extensively there. It has also been commercialized in Germany. The catalyst is the heart of SCR technology, and it must afford both high activity and high selectivity (toward nitrogen formation). Major improvements in SCR catalyst performance can be achieved through strategic design of the catalyst pore structure.

Biodegradation of Organic Waste

Enzymes, like other catalysts, accelerate the rate of reactions. Reactions catalyzed by enzymes include the oxidation and hydrolysis of natural and synthetic organic chemicals regarded as pollutants of soil and groundwater. Enzymes have some natural advantages over other catalysts in the degradation of environmental pollutants. Enzymes are most active against materials at low concentration (micromolar to millimolar range) in the presence of water, and can be simple and inexpensive to manufacture because they are grown along with microorganisms. Enzymes themselves are biodegradable.

Enzymes can exhibit either narrow or broad substrate specificity. The latter characteristic is desirable for enzymes that attack and degrade organic contaminants. For example, methane monooxygenase has an amazingly broad substrate specificity and can catalyze the oxidation of alkenes, ethers, and alicyclic, aromatic, or heterocyclic molecules. This enzyme system can also degrade synthetic organics such as the chlorinated solvents chloroform, dichloroethylene, trichloroethylene, and 1,1,1-trichloroethane. The ability of enzymes to degrade natural organics such as the components of gasoline, crude oil, and most solvents, as well as synthetic organics such as trichloroethylene or polychlorinated biphenyls, means that most, if not all, organic contaminants can be degraded in reactions catalyzed by enzymes.

Chlorinated organics such as dichlorodiphenyl/trichloroethane and its by-product dichlorodiphenyl ethylene, pentachlorophenol, chlorocatechols, and other chlorinated aromatics used as preservatives and pesticides are degraded in oxidation reactions catalyzed by enzymes from bacteria and other microorganisms. Even the most complex halogenated organics, such as polychlorinated biphenyls (PCBs) and chlorinated solvents, are subject to catalytic attack by certain microorganisms.

PCBs were developed earlier in the century as oils for use in electrical equipment and as lubricating fluids in industrial applications because they gave good insulating and lubricating properties without being explosive or flammable. PCBs were later discovered to bioaccumulate and are now classed as environmental hazards. Recently, enzymes have been found in microbes that will reductively dechlorinate PCBs and oxidize them in the presence of molecular oxygen. Even though these enzymes were not evolved to degrade PCBs, they have a broad-enough substrate specificity to catalyze the initial degradation of PCB molecules in the environment. The use of broad substrate specificity oxygenases in bacteria may be the only practical method of treating PCB-contaminated soil and water because of their low cost and adaptability in the environment.

Similarly, chlorinated solvents have been widely adopted because of their excellent solvent properties and lack of flammability. Trichloroethylene

DINING ON POLLUTION

Bacteria and other microorganisms have been digesting other creatures' waste products for millions of years. As a result, there are biological processes to break down most organic compounds, including the amazing variety of complex chemicals synthesized by plants. Some of these processes are now being exploited to break down the complex chemicals made by man. "Bioremediation" is the name given to the use of biological processes to clean up contaminated soil or water.

Almost every organic compound synthesized by man can be broken down, albeit slowly, by microorganisms if their other needs for growth are met. Microorganisms need water, certain inorganic nutrients, and sometimes air to live, but once furnished with those staples, they will obligingly break down many of the synthetic organic chemicals mankind has created, used, and discarded. In fact, scientists have discovered that many toxic chemicals—such as drain cleaners, paint thinners, and used crankcase oil—are routinely broken down by microorganisms that live in wastewater treatment plants downstream.

Enzyme catalysts were used to clean up some of the beaches in Prince William Sound that were contaminated by oil in 1989. Cleanup crews added fertilizer to the beaches, stimulating the growth of their natural populations of microorganisms, which soaked up oil along with the fertilizer and catalyzed its degradation. Without fertilizer to help things along, the microbes would still have digested the oil, but at a very slow rate. The slow rates at which biological processes degrade pollutants such as spilled oil are one reason that bioremediation is not in more widespread use.

However, bioremediation is potentially the most cost-effective approach for cleaning up a contaminated site. Most sites contain a complex mixture of various chemicals. The array of native microorganisms, bolstered when necessary by laboratory cultures of specialized bugs for particularly intractable substances, would simultaneously catalyze the breakdown of the whole assortment of substances present at a site—an approach that could be far cheaper than the current practice of applying a sequence of physical and chemical treatments, each one tailored to particular contaminants. Bioremediation should be less invasive, leaving the soil in place during treatment rather than digging it up, and thus the site should recover more quickly afterward. The process is also self-perpetuating—the bugs will eat the pollutant indefinitely as long as the fertilizer holds out. Thus stimulating natural microorganisms and introducing laboratory-bred or even rationally designed organisms could result in the safe and effective remediation of a large number of contaminated sites across the country.

(TCE), perchloroethylene, and trichloroethane have been used widely in the past as dry cleaning solvents and chemical degreasers for metal finishing and electronic applications. Surprising numbers of different oxidative enzyme systems have been identified that will attack TCE or other chlorinated aliphatics.

The technique of enzyme recruitment offers the prospect of producing single organisms that contain a spectrum of genetically engineered enzymes capable of degrading hazardous waste in the environment. Enzyme recruitment permits microorganisms to degrade new molecules and broadens the ability of microorganisms to attack synthetic organic chemicals. The idea of using enzymes that can reproduce themselves, can be made very cheaply, and can work under conditions that are often found in the environment may be one of the most effective means modern science can devise for treating and degrading hazardous waste, including organic chemicals synthesized by man to be stable under harsh conditions. The broad substrate specificity of certain enzymes offers the opportunity to use enzyme catalysis for improving and protecting the environment.

3

Research Opportunities in Catalytic Science

INTRODUCTION

For viable commercial application, catalysts of any type—heterogeneous, homogeneous, or enzymatic—must exhibit a number of properties, the principal ones being high activity, selectivity, durability, and, in most cases, regenerability. The activity of a catalyst influences the size of the reactor necessary to achieve a given level of conversion of reactants, as well as the amount of catalyst required. The higher the catalyst activity, the smaller are the reactor size and the inventory of catalyst and, hence, the lower are the capital and operating costs. High catalyst activity can also permit less severe operating conditions (e.g., temperature and pressure), and this too can result in savings in capital and operating costs. The amount of reactant required to produce a unit of product, the properties of the product, and the amount of energy required to separate the desired product from reactants and by-products are all governed by catalyst selectivity. As a consequence, catalyst selectivity strongly influences the economics of a process. Catalyst productivity and the time on-stream are dictated by catalyst durability. All catalysts undergo a progressive loss in activity and/or selectivity with time due to chemical poisoning, denaturing, thermal deactivation or decomposition, and physical fouling. When the loss in performance becomes too severe, the catalyst must be either regenerated or replaced. In view of this, high durability and ease of regeneration become important properties.

All of the performance characteristics described above are intimately connected to catalyst structure and composition. It is, therefore, highly desirable to develop a detailed understanding of the relationships among a catalyst's structure, its physical and chemical properties, and its activity, selectivity, and stability. To achieve this goal it is often necessary to characterize catalyst composition and structure at the atomic level. Ultimately, it is most desirable to identify and characterize the catalytic site at the atomic level. Because many catalysts are known to undergo physical and chemical changes under reaction conditions, catalyst characterization should preferably be carried out in situ, or at least under conditions relevant to actual catalytic processes. Knowledge of the interactions between a catalyst and reactants, intermediates, and products is needed to understand the influence of the catalyst on the structure and bonding of species involved in catalyzed reactions. The dynamics of chemical transformations occurring under the influence of the catalyst is yet another area in which information is needed.

The development of new or improved catalysts is complex, involving extensive testing and evaluation. Because the range of variables is often very large, and the relationships between changes in these variables and catalyst performance are often not clearly identified, catalyst development can be a tedious and expensive task. Knowledge derived from scientific studies provides a basis for conceiving new catalysts and catalytic reactions, and for interpreting the results of experimental observation. Moreover, many of the analytical techniques developed in the pursuit of catalytic science can be used effectively to elucidate cause-effect relationships and, thereby, to accelerate the process of catalyst development.

Opportunities for advancing the frontiers of catalytic science exist in four areas: the synthesis of new classes of catalytic materials, catalyst characterization, the mechanism and dynamics of catalytic reactions, and the theory of catalysis. Each of these areas is highlighted below, with indications given for future research directions.

SYNTHESIS OF CATALYTIC MATERIALS

There are three reasons for pursuing research on the synthesis of catalytic materials. The first is to find new or improved catalysts for a desired reaction (e.g., efficient production of high-quality fuels, the decomposition of nitric oxide, the direct conversion of methane to methanol, the synthesis of homochiral or enantiomerically pure drugs). In this instance, either new classes of materials or modifications of existing materials are sought to achieve the desired increase in activity or selectivity. The second reason for studying catalyst synthesis is to establish the relationships between preparative procedure and final catalyst structure and properties. The objective in this case is to understand how the choice of starting materials and synthesis conditions influences catalyst composition and structure. Success in this endeavor can lead to the identification of principles and strategies for preparing catalysts with specified properties. The third reason is to reduce manufacturing costs by reducing raw material or processing costs.

The importance of catalyst synthesis is well illustrated by the recent development of high-activity catalysts for the reduction of nitrogen oxide (NO_x) emissions from power plants. The material of choice for this application is titania-supported vanadia. In currently practiced technology, vanadia is dispersed into the pores of a titania monolith. A reaction engineering analysis of the performance of such catalysts has revealed that the catalyst operates in the diffusion-influenced regime and that the pore structure of the support can be optimized for maximum performance. Further research has shown that the desired pore structure cannot be achieved by using bulk titania because of physical strength constraints, but can be achieved by using silica. To obtain the chemical properties of titania required for high intrinsic activity, titania is dispersed into the pores of a silica monolith, and vanadia is then deposited on the titania particles. Based on laboratory-scale tests, the resulting material exhibits a catalytic activity 50% higher than that previously available and promises improved poison resistance due to its bimodal pore structure. This illustration shows the manner in which knowledge of material properties can be combined with an analysis of reaction dynamics and mass transfer to design a catalyst with optimal performance characteristics for a targeted application.

High selectivity in combination with high activity can often be achieved with homogeneous catalysts. These properties are influenced by the nature of the transition metal situated at the catalytically active center of the complex. Variations in the composition of the ligands and the solvent in which the complex is dissolved can influence the catalytic properties of the complex. Strategic manipulation of these variables can be used to obtain useful catalysts. A recent illustration of this point is the synthesis of naproxen, an anti-inflammatory drug. As currently manufactured, this drug is expensive because the synthesis procedure results in a mixture of the two optical isomers that must be separated, because S-naproxen is the desired product but R-naproxen is a liver toxin. To reduce the costs of production, one wants to increase catalyst selectivity for the S-isomer. Recent research has shown that naproxen can be produced with high selectivity by asymmetric hydrogenation of α-(6-methoxy-2-naphthyl) acrylic acid using a soluble ruthenium complex containing a chiral phosphine ligand. This advance holds promise not only for reducing the cost of production, but also for eliminating potentially harmful by-products.

Molecular sieves, of which zeolites are a special class, offer extensive opportunities for the design of new catalysts. These materials are characterized by a crystalline framework containing cavities and channels of molecular dimensions (0.3-1.0 nm). Catalytic activity is typically due to acidic sites in the framework. Multifunctional catalysis can be achieved by dispersing small metal particles within the cavities of the molecular sieve.

What makes molecular sieves such interesting materials is their wide range of compositions and topologies. Prior to 1982, most molecular sieves were based on aluminosilicates. These materials are commonly known as zeolites. However, recent developments at Universal Oil Products (UOP) Inc. have demonstrated the feasibility of producing molecular sieves based on aluminophosphates. By substitution of various metallic elements for either aluminum or phosphorus, a broad range of materials can be produced that vary in their catalytic properties. Although about 60 different molecular sieve topologies are known today, tens of thousands of structures are theoretically possible, but knowledge of how to synthesize specific structures is not yet available. An issue of particular interest is to find ways of producing molecular sieves with openings in excess of 0.7 nm. Such materials would be able to accept the large reactant molecules found in heavy petroleum and in other heavy fossil liquids. Progress toward this goal has recently been achieved by the discovery of a large-pore molecular sieve (VPI-5) that contains 18-membered oxygen rings with an opening of 1.2 nm. The challenge now is to find ways to make VPI-5 catalytically active and stable under reaction conditions.

Molecular sieve crystals provide not only uniform intracrystalline pores, but also uniform and well-defined tetrahedral coordination for framework metal ions. The discovery of a great variety of structures and chemical compositions in the aluminophosphate molecular sieve family of crystals opens opportunities to prepare molecular sieves with transition metal cations in the crystal framework. This unique, predictable positioning of transition metal ions in microporous crystal lattices offers opportunities in oxidation, redox, or other transition-metal-related catalysis.

Clays, phosphonates, and other lamellar structures represent yet another class of materials that are of growing interest to the field of catalysis. By introducing either organic or inorganic "pillars" between the lamellae, it is possible to create galleries of molecular dimensions. Through the control of pillar height, uniform interlamellar spacings of 0.5-4.0 nm can be achieved. Such pillared layered structures can be used as catalysts if the pillars contain acidic groups or as supports for transition metals or their complexes. In the latter case, the support can serve to orient the reactant so that unique reactant and product selectivities can be achieved. A significant challenge for the future is to find ways to improve the thermal and hydrothermal stability of such pillared lamellar structures.

Enzymes are highly efficient catalysts that have great potential industrially, especially for the synthesis of chiral compounds. Until the advent of recombinant DNA technology, this potential could not be realized. Previously, only those few enzymes that were obtainable in large amounts from

THE POLYETHYLENE STORY: CUT-RESISTANT SURGICAL GLOVES AND LAUNDRY BAGS

What do cut-resistant surgical gloves and laundry bags have in common? They are both made from polyethylene. Polyethylene was discovered in the late 1930s by a group of chemists who were studying the properties of ethylene, a gas, at high pressure and temperature. They discovered that under certain conditions, the gas turned into a flexible white solid that proved to be an excellent insulator. (It has since found favor as a food wrapper.) World War II created an urgent demand for large quantities of such an insulator to coat the wires of radar sets and electronic weaponry. Without that demand, in fact, polyethylene might have been abandoned as too dangerous to manufacture: the early production plants often blew up.

As it turned out, World War II also resulted in a safer way to make polyethylene. Germany, cut off from the world's oil fields by the Allied blockade, mounted a crash research program to develop synthetic fuel and lubricating oils. Karl Ziegler, one of the chief chemists in that effort, continued this research after the war, using ethylene as his raw material. One day in 1953, he was surprised to find the reaction vessel full of polyethylene, although he had not been using the high temperature and pressure then thought necessary to make it. The secret proved to be a metal catalyst. Unlike the high-temperature variety, this new polyethylene was rigid, not flexible, and it differed in other ways as well. (It is now used to make milk bottles.)

Further research showed that the two types of polyethylene grow in very different ways. Ziegler's catalyst selectively stitches ethylene molecules together end-to-end to make long, straight chains of polyethylene. These linear molecules stack readily, like cordwood, to give the rigid, crystalline, high-density material that is used for milk bottles. The high-temperature process, however, proceeds through "radical" intermediates—unstable species that react indiscriminately with anything in sight, including their own polymer chains. When this happens, the once-linear chain develops branches and, just as a pile of brushwood does not stack neatly, so neither do the branched polymers, which results in a soft, amorphous, low-density material.

Modern catalysts can be used to polymerize ethylene into an ultra-high-molecular-weight linear polymer. This material can be spun into extremely strong fibers that can be used in cut-resistant surgical gloves.

Today's catalysts have been much improved since Ziegler's discovery, so that very selective processes now produce very specific products in very high yields. Further improvements will result not only in milder processing conditions but also in superior material properties for a broader range of uses.

natural sources could be considered for commercialization. Even some natural enzymes had properties not directly suitable for a particular industrial application. Now it is possible to produce any enzyme in industrially useful form by cloning the gene and expressing it in a suitable production organism. The properties of the enzyme, such as stability, selectivity, and catalytic efficiency, are controlled by its three-dimensional structure. This structure is in turn determined by the sequence of amino acids in the enzyme, which can now be altered by changing the gene that codes for the enzyme. The researcher can fit the desired application by changing the amino acid sequence in the enzyme. This has been done for several enzymes in commercial use and will lead to the increased use of enzymes as catalysts for industrial processes.

Another area of rapidly emerging opportunity is the formation of catalytically active antibodies targeted for specific reactions. Various approaches are available. These include the raising of antibodies to structures resembling those of the transition states of the reactions for which a catalyst is desired. Recent studies have demonstrated rate accelerations for a number of antibody-catalyzed reactions of up to 10^6 times the uncatalyzed rate. A development in the design of catalytic antibodies involves the introduction of catalytic activity into antibodies via either molecular biological or chemical means. These strategies allow for the a priori evolution of catalytic activity in an antibody combining site and also allow incremental increases in the rates of catalytic antibodies generated by other means. Examples of such approaches include introduction of catalytic groups via hapten-antibody complementarity, chemical modification, generation of cofactor binding sites, and site-directed mutagenesis. These systems offer catalysts for reactions beyond those available from enzymes.

Within the United States, the design and synthesis of new catalytic materials have been pursued largely in industry but have received much less attention in government and university laboratories. This is not the case in Japan and the European Community, where considerable research on catalyst synthesis is evident, and in fact targeted, in many academic research institutions. The absence of a strong effort toward the synthesis and evaluation of all types of novel materials is a weakness in the catalysis program in the United States. Not only does this sharply limit the techniques that can be brought to bear on problems in this area, but it also curbs the opportunities for discovery.

As illustrated by the examples presented above, many exciting research opportunities are available for the design and synthesis of all types of new catalytic materials, and strong support should be given to this important area.

CATALYST CHARACTERIZATION

Knowledge of catalyst composition and structure is critical to a fundamental understanding of the chemistry actually occurring in catalysis. From such information it is possible to determine which portions of a catalyst are active and how changes in catalyst synthesis and structure affect the properties of such sites. Catalyst characterization is also vital to understanding the changes that occur in the structure and composition of a catalyst following pretreatment, during induction periods, in use under reaction conditions, and during regeneration.

Over the past two decades, advances in analytical instrumentation have led to significant improvements in sensitivity and resolution. This has enabled researchers to obtain detailed structural and compositional analyses, often with atomic spatial resolution. Techniques currently available for catalyst characterization are listed in Table 3.1.

Table 3.1 Experimental Techniques for Characterizing Catalysts and Adsorbed Species

Technique	Acronym	Type of Information
Low-energy electron diffraction	LEED	Two-dimensional structure and registry with metal surface
Auger electron spectroscopy	AES	Elemental analysis
X-ray photoelectron spectroscopy	XPS	Elemental analysis and valence state
Ion scattering spectroscopy	ISS	Elemental analysis
Ultraviolet photoelectron spectroscopy	UPS	Electronic structure
Electron energy loss spectroscopy	EELS	Molecular structure
Infrared spectroscopy	IRS	Molecular structure
Laser Raman spectroscopy	LRS	Molecular structure
X-ray diffraction	XRD	Bulk crystal structure
Extended x-ray absorption fine structure	EXAFS	Bond distance and coordination number
Transmission electron microscopy	TEM	Crystal size, shape, morphology, and structure
Scanning transmission electron microscopy	STEM	Microstructure and composition
Scanning tunneling microscopy	STM	Microstructure
Ultraviolet spectroscopy	—	Electronic state
Mössbauer spectroscopy	—	Ionic state
Nuclear magnetic resonance spectroscopy	NMR	Molecular structure and motion

STUDIES OF AMMONIA SYNTHESIS OVER
IRON SINGLE CRYSTALS

Since 1913, ammonia has been produced in tonnage quantities by direct combination of nitrogen and hydrogen at high temperature and pressure. Most of it is used to make nitrogen-rich fertilizers, and the balance to manufacture explosives, of use in peace as in war. Ammonia synthesis occurs at the surface of iron catalysts. Because ammonia is a low-priced commodity chemical, the catalyst must be cheap and durable, and its activity as high as possible, so that temperature and pressure can be kept low to minimize the size and cost of huge industrial reactors. Factors responsible for the activity and durability of commercial iron catalysts have been elucidated in the 1980s by surface science studies conducted with model catalysts consisting of single crystals of iron approximately 1 cm in size.

The rate of ammonia synthesis at high temperature and pressure was measured over five of these large crystals of pure iron, cut in five different ways so as to expose five different facets. Two facets were found to be almost equally active, but more active in producing ammonia than the other three. It is logical to expect that the best commercial iron catalysts should expose one or both of the best facets. Is it so?

The answer seems to be yes, for a very simple reason. When the activity of a commercial iron catalyst is compared to that of single crystals under identical conditions, it is found that both values per iron atom exposed are about the same and equal to those found on the best two facets. Thus, it appears that the commercial catalyst has been optimized. How did this happen?

The answer to this second question is more subtle and related to the other desirable feature of a catalyst, namely, durability. Commercial iron catalysts consist of very small crystals of iron, a few millionths of a centimeter in size. As a result, the catalyst exhibits a very large surface area per unit volume, a most desirable characteristic to minimize reactor size.

However, tiny crystals of pure iron fuse under the harsh temperatures of ammonia synthesis, a phenomenon that leads to catalyst sintering and the death of the catalyst. Indeed, a commercial iron catalyst is not made of pure iron metal, but of iron to which certain oxides have been added. Because the added oxides are beneficial, they are called promoters. One of these promoters is aluminum oxide, alumina. Alumina by itself or on an iron surface is totally inactive in ammonia synthesis. Its role as a promoter has been ascribed in the past to its ability to prevent sintering of iron particles (i.e., to ensure catalyst durability).

Recent work with large single crystals does not refute this interpretation, but adds a significant dimension to the role of alumina as a promoter. Thus, surface science studies have revealed that the addition of alumina to the most inactive facet of a large iron single crystal, followed by heating under appropriate conditions, causes its recrystallization to form ultimately the two most active facets during ammonia synthesis. The conclusion is that alumina brings about the restructuring of any iron crystal face to produce and maintain those facets that are most active for ammonia synthesis.

It has been said that surface science work with large metallic single crystals provides the standards by which work with commercial metallic catalysts can be assessed and understood. This vignette provides a striking illustration of that statement.

Experience has demonstrated that no single technique can provide a complete picture but that, instead, a combination of techniques is necessary. The characterization of zeolites is a case in point. The lattice structure of a zeolite can be determined from x-ray diffraction by using either single-crystal or powder-diffraction techniques. The correctness of the lattice structure may be confirmed by atomic resolution electron microscopy, which also provides evidence of lattice defects or of the presence of minority impurity phases. The distribution of the elements constituting the lattice framework can be established by ^{29}Si magic angle spinning (MAS) nuclear magnetic resonance (NMR) spectroscopy, and the nature and strength of Brønsted acid sites can be determined by using ^1H NMR spectroscopy or a variety of sorption techniques. The size, crystallographic perfection, and defects of cavities inside the zeolite can be probed by ^{129}Xe NMR spectroscopy. If metal particles are occluded within the zeolite cavities, the average size of these particles can be established by extended x-ray absorption fine structure (EXAFS). By combining information from the above techniques, it is possible to develop a detailed atomic view of a zeolite catalyst that can serve as the starting point for developing theoretical models of catalytic sites (see below) and identifying strategies to improve catalyst performance.

Acid catalysis applied both in homogeneous solution and over solid catalysts is the largest branch of industrial catalysis. In homogeneous solution the understanding of reaction mechanisms is aided greatly by the broad knowledge of the chemistry of homogeneous acid solutions. Understanding the chemistry over solid acids is hampered by a lack of absolute acidity measurements or of sound comparisons between liquid and solid acids. The recent application of ^1H NMR in combination with other methods promises to provide an absolute scale and measurement of the acid strength of solid acid catalysts. If successfully developed, such techniques could be applied to the development of a variety of practical solid acid catalysts.

Heterogeneous noble metal catalysts are generally prepared by dispersing the metal in the form of small crystallites onto the surface of a metal oxide such as alumina or high-surface-area porous carbon. The size and morphology of metal particles can be established from transmission electron microscopy. Electron microscopy can also provide information on the nature of the interaction between the metal particles and the support. EXAFS can be used to determine the average metal-metal bond length and coordination number. For bimetallic catalysts (e.g., platinum-iridium, platinum-rhenium), the degree of interaction and the distribution of the two components can also be established.

A large body of evidence acquired in recent years has shown that the properties of metal catalysts can be modeled by using single-crystal metal surfaces. This method has proved invaluable for the advancement of catalyst science because of the large array of techniques developed by surface scientists for the characterization of single-crystal surfaces. The atomic composition of such surfaces can be analyzed by using Auger electron spectroscopy (AES), x-ray photoelectron spectroscopy (XPS), and ion scattering spectroscopy (ISS). The structure of the exposed surfaces and the presence of terraces and defects can be established by low-energy electron diffraction (LEED). An exciting recent development, scanning tunneling microscopy (STM), has made it possible to obtain atomic-resolution views of surfaces and to identify the influence of thermal and chemical pretreatment on surface structure.

Insights into structure-function relationships can be obtained through the use of model catalyst systems. Examples of such systems include metals deposited on oxides, oxides on oxides, and oxides on metals. After initial preparation by means of vapor deposition, the sample can be characterized by AES and XPS. The catalytic properties of the catalyst can be determined by transferring the sample to a high-pressure chamber. Characterization of the sample after reaction reveals changes that have occurred in the catalyst composition and structure. Model catalyst systems are also well suited for examination by means of scanning transmission spectroscopy (STEM) and controlled atmosphere electron microscopy (CAEM). These techniques can provide detailed evidence for surface reconstruction, sintering, alloying, and phase separation.

Significant advances have also been made in the characterization of enzymes, catalytic antibodies, and homogeneous catalysts. Where such catalysts can be obtained in the form of single crystals, the analysis of x-ray diffraction patterns can provide a complete atomic structure. However, although such information is quite useful, it does not indicate structural changes that may occur when the catalyst is saturated and surrounded by solvent. Recent advances in multidimensional (i.e., two-, three-, and four-dimensional) NMR spectroscopy are making it possible to obtain this information in solution, and enzyme structures containing as many as 153 amino acids have been analyzed completely.

The detailed mechanism of some homogeneous transition metal catalysts has been elucidated by work in the closely allied field of organometallic chemistry. In many cases, intermediates can be trapped and models for catalytically active species can be prepared and studied for kinetic competence in the catalytic process. These complexes can be characterized completely by small-molecule crystallography and examined in solution by high-field NMR techniques.

Because the advancement of catalysis along the experimental front depends heavily on the availability of advanced instrumental characterization techniques, it is important that research on the development of new techniques be well supported. The United States has traditionally been a leader in the field of instrumental analysis, and many of the techniques in use today originated in academic laboratories. This position is currently in jeopardy, because of strong competition from the European Community and Japan. Particular emphasis should be placed on the development of techniques that can be used to characterize catalysts in situ. Examples of such techniques include STM, atomic force microscopy (AFM), NMR spectroscopy, and EXAFS.

MECHANISM AND DYNAMICS OF CATALYTIC REACTIONS

Knowledge of catalytic reaction mechanisms—in particular the structure, dynamics, and energetics of reaction intermediates formed along the catalytic reaction path—can provide insights for developing new catalysts and improving existing ones. In situ observation, in turn, is the most effective technique for elucidating reaction mechanisms, the dynamics of molecular interactions with catalysts, and the structure of stable intermediates. Such investigations also provide information on the thermodynamics of catalyst-substrate interactions and the associated activation barriers. Homogeneous catalysts can also be studied in solution by NMR techniques; these studies have resulted in new homogeneous catalytic reactions, and they provide models for related heterogeneous processes. Several examples will serve to illustrate the importance of a multitechnique approach.

L-Dopa, used in the treatment of Parkinson's disease, is prepared by asymmetric hydrogenation. The catalysts used in this process are homogeneous rhodium complexes. When synthesized with chiral phosphine ligands, these complexes enable the production of the desired enantiomer with greater than 95% selectivity. Detailed mechanistic studies involving NMR spectroscopy have permitted the identification of catalytic intermediates and have led to the design of new asymmetric hydrogenation catalysts and to the selection of optimal reaction conditions. As the emphasis on producing optically pure drugs increases, determination of the details of homogeneous asymmetric catalysts will become increasingly important.

CONTROLLING BUGS WITH BUGS

Are there ways to control gypsy moths, cabbage loopers, and other agricultural annoyances without using chemical pesticides? The answer is yes. There are now available a number of biological products registered with the Environmental Protection Agency (EPA) that can be used to control a fairly wide array of insects, weeds, and other plant pests (Figure 3.1). Most of these products are based on microorganisms ("bugs") that are the pest's natural enemies—disease-causing "pathogens" that kill the pest outright or foul-smelling, worse-tasting "antagonists" that drive the pest away.

These biopesticides have several advantages over chemical pesticides. Biopesticides act through mechanisms specific to the target organism and nonexistent in other organisms, thus sabotaging the target's metabolism without poisoning every other creature as well. Biopesticides are fully biodegradable, leaving no residues behind, and biopesticides can be applied with the same equipment used to apply conventional pesticides, at the same intervals, and with essentially equivalent results.

If these marvelous products are truly available, why isn't there a greater awareness of them? There are several reasons. Although some of these products have been around for 20 years or more, they did not work as well in the past as they do now, thanks to continuing research. Nor have they ever been price competitive with conventional pesticides. Where they have been used extensively, for example, in eastern U.S. oak forests to control gypsy moths, they have been well received. They have, however, generally been seen as specialty products for "niche" markets, rather than as alternatives to chemical pesticides in the broadest sense.

This view may be changing. Opposition to chemical pesticides continues to increase with increasing awareness of their environmental consequences. The cost of developing new chemical pesticides and of registering them with the EPA—a precondition to their sale in the United States that requires extensive environmental and toxicological studies— has led to a decline in the rate of new product development. Furthermore, many insects and plant diseases have shown an amazing ability to develop resistance to an ever-broadening array of pesticides, limiting the useful life of even newly developed products. Biopesticides, on the other hand, have received increasing attention over the past five years. Several new biotechnology companies are focusing specifically on biopesticide development, and even old-line agricultural chemical companies have begun to shift their pesticide lines toward environmentally compatible products.

Biotechnology—biocatalysis—is indispensable to biopesticide development, as the story of the development of insecticides from the soil bacterium *Bacillus thuringiensis* shows. Over the past 10 years, scientists have found that this bug's insecticidal properties are encoded by a diverse array of genes that direct the synthesis of insecticidal proteins. These proteins act as selective poisons in the stomachs of insect larvae. Different genes produce poisons specific to various plant-attacking beetle larvae and caterpillars, as well as to the larvae of such carriers of human disease as mosquitoes and blackflies. By isolating and analyzing the genes involved, scientists can now "engineer" new gene combinations that improve on their naturally occurring counterparts, creating more potent weapons against a broader range of pests. Such engineering is an exquisite example of biocatalytic synthesis, using the bacterium's metabolic machinery to create proteins that would be prohibitively expensive to manufacture by industrial chemical processes. The new genes can be transferred to other bacterial hosts. Plants can also be "vaccinated" by endowing other organisms that grow in or on them with the gene. This scheme, which would provide a steady supply of the biopesticide to the plant, may prove more efficient than conventional spraying.

Those who protest the continued use of pesticides should be gladdened to know that there is hope for alternative products. Many are available now. Others will become available in the near future, if industry continues to support their development. For these new products to make their contribution, however, regulatory agencies and the public must be willing to agree that genetically engineered products are safe for use in the environment. The benefit to be derived from these products must be shown clearly to outweigh any perceived risk. This is the challenge to be faced if we are to continue to maintain and increase our agricultural productivity.

To produce high-octane components for gasoline, normal paraffins must be converted into branched paraffins, olefins, and aromatic compounds. Such transformations can be carried out over a platinum catalyst under a hydrogen atmosphere in a process called reforming. Extensive efforts have been undertaken to understand the fundamental chemistry involved in the reforming process. Vibrational spectroscopies, such as electron energy loss spectroscopy (EELS) and infrared spectroscopy, and ^{13}C NMR spectroscopy have revealed that paraffins are only weakly adsorbed but undergo a loss of hydrogen on heating. The exact amount of hydrogen lost can be determined from temperature-programmed desorption spectroscopy, whereas the structure of the surface species formed as the temperature increases can be determined from both infrared and NMR spectroscopies. These techniques are particularly valuable because they can be used at el-

Figure 3.1   Effects of the biopesticide Foil on the growth of potatoes.   Shown are an untreated plot (top) and a treated plot (bottom).   (Figure courtesy of Ecogen.)

evated temperature and pressure. Additional information on the structure of adsorbed hydrocarbons has been obtained from LEED studies. More recently, molecular beam investigations have proved useful in determining the dynamics of hydrocarbon adsorption and decomposition on metal surfaces. Because the thermal energy of the hydrocarbon can be controlled in such experiments, it is possible to determine the kinetics of hydrocarbon reactions in great detail. Studies carried out with well-defined single-crystal surfaces of platinum have demonstrated the sensitivity of reaction dynamics to the crystallographic planes exposed and to the presence of defects such as steps and kinks.

Ethylene oxide and propylene oxide are critical intermediates in chemical manufacture. Ethylene can be catalytically oxidized by molecular oxygen over a silver catalyst to give an excellent yield of ethylene oxide. The synthesis of propylene oxide requires the catalytic transfer of oxygen from an alkyl peroxide to propylene. Transition metal catalysts, usually based on molybdenum, are essential for the reaction. These two reactions have a major impact on the commodity chemical industry. Recently, the synthesis of epoxides has been modified to yield enantiomerically pure products, an advance that has had a major impact on the manufacture and synthesis of bioactive molecules. Barry Sharpless' group, at MIT, has shown that titanium complexed to tartrate esters (obtained from nature as one optical isomer) will transfer oxygen from tertiary-butyl hydroperoxide to allyl alcohols to give enantiomerically pure epoxides. Such studies have provided the paradigm for the role of structure, mechanism, and dynamics in the discovery of selective catalysts for the production of enantiomerically pure products.

The conversion of methanol to gasoline over ZSM-5, a medium-pore zeolite, has also been investigated extensively as part of the effort to develop alternatives to petroleum as a source of transportation fuels. Both ^1H and ^{13}C NMR have been useful techniques for characterizing the intermediates formed between olefins and protic sites inside the zeolite and for characterizing the structure of coke, which builds up slowly with time and contributes to catalyst deactivation. This type of information has proved useful in understanding how zeolite acidity and pore size influence the rate at which deactivation occurs.

The selective catalytic reduction of nitric oxide by ammonia over titania-supported vanadia catalysts provides an effective means for reducing NO_x emissions from stationary sources. Isotopic tracers have revealed that this reaction is initiated by the formation of an adduct between nitric oxide and ammonia, whereas Raman and NMR studies of adsorbed ammonia have shown how ammonia complexes with vanadium ions present on the catalyst surface. Such investigations suggest that oxygen atoms bound to vanadium ions play an important role in activating the adsorbed ammonia for reaction. The role of additive oxides that are known to enhance catalyst activity and selectivity can be investigated by using spectroscopic techniques. Such promoters interact with the reactants and suppress the oxidation of sulfur dioxide to sulfur trioxide, an undesired side reaction.

In summary, investigations of reaction mechanisms and kinetics and, especially, in situ observations of catalytic reaction intermediates are essential for advancing the science of catalysis, inasmuch as the results of such studies provide an overall view of catalysis and help elucidate the relationships between catalyst structure and function. The current interest in developing catalysts for the production of environmentally benign gasoline, the abatement of air pollution from mobile and stationary sources, the synthesis of enantiomerically pure drugs, and the synthesis of novel polymers all benefit from studies of the relevant reaction mechanisms and kinetics. The successful advancement of knowledge in this area requires the further development of techniques for characterizing adsorbed species, particularly their structure and bonding; for identifying the connectivity between species in terms of a reaction network; and for characterizing the dynamics of elementary chemical transformations occurring under the influence of the catalyst.

THEORY OF CATALYSIS

The science of catalysis has traditionally advanced as a consequence of new experimental techniques, but theory has begun to play an increasingly important role. This change is the result of recent advances that have occurred in theoretical and computational chemistry, reaction engineering, and the availability of powerful supercomputers for extensive calculations, which enable graphical display of results. Information gained from theoretical studies is becoming helpful in guiding the design of novel catalysts, interpreting experimental measurements, and understanding the way in which catalyst composition and structure affect its activity and selectivity.

Considerable progress has been made in modeling and calculating the relative energies of intermediates in homogeneous transition metal catalysis. These systems are small and generally involve one metal and one constant ligand set. However, even for relatively simple systems, major approximations are required for ab initio calculations. The results obtained from these systems in which detailed molecular structures can be determined and systematic structural changes can be made will serve as guides for the more complex heterogeneous systems.

Theoretical calculations of catalyst structure can provide a useful basis for assessing the stability of particular structures as a function of composition and surrounding environment. For example, in the area of zeolite catalysis, it is possible to predict the stability of channel openings as a function of the ratio of silicon (Si) to aluminum (Al) and the preferred location of aluminum in the framework. Similarly, theoretical models of

H+
|
Si–O–Al suggest a direct relationship between Si–O–Al bond angles and zeolite acid strength, consistent with experimental observation. In the case of bimetallic catalysts, the spatial distribution of each component in small particles can be computed for a given particle size and composition. Computational chemistry has also proved useful in examining the influence of site-specific modifications and solvent effects on the structure of active sites in enzymes.

The interpretation of spectroscopic analyses of catalysts can be facilitated with the aid of appropriate computations. Calculations of the occupation of valence electrons in d-orbitals have proved useful in the proper interpretation of x-ray absorption near-edge structure (XANES) measurements on small metal particles. For zeolitic materials, theoretical predictions of infrared can be made. Comparison of calculated and experimentally observed spectra can be used as a sensitive test of atom-atom potentials. With accurate knowledge of such potentials, it is possible to make predictions about phase transformations and other structural changes occurring at elevated temperature.

There are an increasing number of examples in the literature indicating the feasibility of computing the potential energy curves for elementary steps in catalyzed reactions of small molecules. Figure 3.2 illustrates a potential energy diagram obtained from an ab initio quantum chemical study of ethylene hydrogenation by using a model Wilkinson catalyst, $RhCl(PH_3)_3$. Such calculations are useful, as a first step, for identifying the most energetically demanding (i.e., rate-limiting) steps in a reaction sequence; however, future efforts along these lines must be carried out by using realistic phosphine ligands so that meaningful comparisons can be made with experimental observations. Prediction of the influence of solvents on the energetics and dynamics of homogeneous catalytic reactions also remains an important challenge. Some initial attempts to address such problems have been carried out in the area of enzyme catalysis, where the properties of the solvent are known to influence the rate of substrate-enzyme association.

Progress has also been made in evaluating the energetics of atoms and small molecules interacting with metal surfaces. Atomic heats of adsorption can now be calculated by using ab initio techniques, which are in very good agreement with experiment. Studies of molecular adsorption have identified the preferred sites for adsorption and the structure and bonding of the adsorbed molecule. To date, however, ab initio techniques have proved too expensive to permit full exploration of the multidimensional potential surface governing reactions. As a consequence, less accurate, semiempirical methods have been explored. Such techniques have provided information regarding the dissociation path of diatomic molecules on transition metal surfaces and the activation barriers for various elementary reactions. Cal-

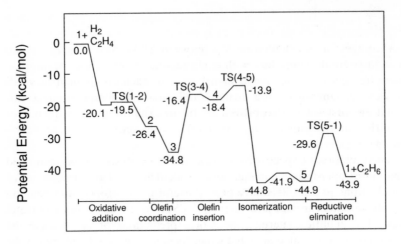

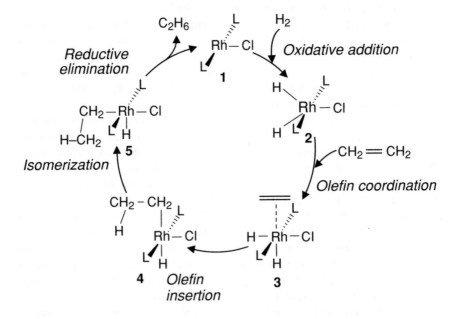

Figure 3.2 Potential energy diagram obtained from an ab initio quantum chemical study of ethylene hydrogenation by using a model Wilkinson catalyst. (Reprinted, by permission, from N. Koga, C. Daniel, J. Han, X. Y. Fu, and K. Morokuma, 1987, pp. 3455–3456 in *Journal of the American Chemical Society*, Vol. 109. Copyright © 1987 by the American Chemical Society.)

culations of the effect of poisons (e.g., sulfur) and promoters (e.g., K^+) have yielded information on the manner in which the components affect the adsorption of molecules. Computed potential energy surfaces can be used to calculate the rate coefficients for elementary processes such as adsorption, diffusion, desorption, and reaction. By use of dynamic simulation techniques and dynamically corrected transition state theory, estimates of rate parameters have been obtained that are in reasonable agreement with experimental observation.

The application of theoretical methods to zeolites has proved fruitful. Advances in theory have contributed to understanding the thermodynamics of adsorption in zeolites and the dynamics of diffusion. Monte Carlo calculations can now provide accurate predictions of the isotherms and heats of adsorption of hydrocarbons in a number of zeolites. Molecular dynamics calculations can be used to describe the motion of molecules through the void space of zeolites. Such calculations have been used to determine the molecular diffusion coefficients, various properties of motion, and the spatial distribution of sorbates within the zeolite void space. Promising progress has also been made recently in the use of ab initio quantum chemical techniques to characterize the interaction of reactant molecules with acid sites within a zeolite. Such calculations can provide a description of the reactant-zeolite potential surface, from which it will then be possible to define the reaction intermediates and to determine the rate coefficients for chemical transformations.

Yet another area in which progress is being made is the modeling of catalyst particle and reactor performance. Such models combine information about intrinsic catalyst activity and selectivity, intraparticle mass and heat transfer, and heat, mass, and fluid transport within the reactor to predict product conversion and yield, and catalyst performance with time onstream. Models of this type can be used to determine the extent to which transport phenomena affect catalyst or reactor performance and the mode by which catalyst deactivation occurs (e.g., sintering, poisoning, fouling).

It has also become possible to use catalyst and reactor models to determine the optimal design of catalyst particles and structures. In such cases, calculations are made to maximize or minimize a desired objective function for a given set of design and operating parameters. Typical catalyst design parameters include descriptors of the structure and composition of the catalyst surface, the pore structure, the activity distribution in the catalyst particle, and the size and shape of the catalyst particle. Typical operating parameters are reactor type (fixed bed, ebullated bed, fluid bed, and so on), inlet conditions (temperature, pressure, composition, flow rate), and parameters related to heat or solid flow management. The objective function (or functions) to be maximized (or minimized) provides a quantitative measure by which the performance of the catalyst, catalytic reactor, or catalytic process is to be judged

SOLID-STATE HIGH-TEMPERATURE FUEL CELLS

An anonymous wit once paraphrased the three laws of thermodynamics thus: First, you can never win, you can only break even. Second, you can never break even. Third, you can't get out of the game. The second law limits the efficiency of a fossil-fuel-fired power plant—only a fraction (typically less than 40%) of the chemical energy released by burning the fuel is converted to electricity. The balance is dissipated in various unavoidable ways—as friction between moving parts, as waste heat up smokestacks and cooling towers, and so forth. Fuel cells, however, translate chemical energy directly into electrical energy without any mechanical or thermal intermediaries. Fuel cells can have efficiencies as high as 90%, depending on their applications. With such high efficiencies, power plants based on fuel cell technology would consume much less fossil fuel and emit proportionally fewer pollutants than would conventional power plants.

All fuels burn by reacting with oxygen to release energy, and the key to a fuel cell's efficiency lies in using catalysts to control that reaction. There are a number of different types of fuel cells. One promising variety is the high-temperature, solid-state fuel cell (Figure 3.3), which is essentially a barrier made of ceramic, typically zirconium oxide doped with traces of yttrium oxide, whose structure conducts oxygen ions (negatively charged oxygen atoms). On one side of the barrier—the fuel cell's negative terminal—the fuel, typically a mixture of carbon monoxide and hydrogen gas created by gasifying coal or steam-refining natural gas, reacts catalytically with oxygen ions to liberate water, carbon dioxide, and electrons. The electrons go out over the wire as an electric current. On the other side of the barrier—the positive terminal—the electrons returning through the wire are catalytically added to molecules of oxygen from the atmosphere to create more oxygen ions that diffuse through the barrier and perpetuate the cycle. The yttria-doped zirconia fuel cell typically operates at around 1000° C.

So what are we waiting for? Why aren't power plants based on fuel cells? Unfortunately, several technological hurdles must be overcome before this catalytic technology can be a commercial success. A worldwide effort has been under way for more than two decades to clear these hurdles, which include devising ways to keep the catalyst from breaking down at such high temperatures, avoiding cracks and leaks in the ceramic structure, and designing a ceramic that conducts enough oxygen ions through a sufficiently small volume to make the size of the equipment economical. Some current efforts are focused on so-called cross-flow monolithic designs in which a solid ceramic block is laced with sets of fuel channels perpendicular to, and alternating with, sets of air channels. This design provides a high interior surface area upon which reactions can occur and thus occupies a very compact volume per unit of energy generated.

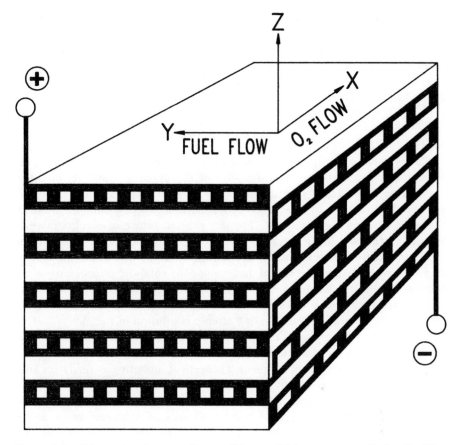

Figure 3.3 Schematic of a cross-flow, solid-state, high-temperature fuel cell. (Figure courtesy of W. R. Grace & Company.)

(e.g., maximum feed conversion, maximum time on-stream, minimum coke loading). As advances are made in the theoretical prediction of transport and reaction in catalysts, one can envision the development of a hierarchy of algorithms for describing catalyst performance, starting at the molecular level and progressing upward to the level of a complete reactor.

In summary, the application of theoretical methods to describe and predict the properties of catalysts is a rapidly developing area of research that can contribute much to the scientific understanding of catalysts and the chemical reactions occurring under their influence. Critical areas for future research include the computation of accurate catalyst-substrate potential surfaces by using ab initio quantum chemical techniques, the evaluation of rate and diffusion coefficients via molecular dynamics and transition state theory, the application of Monte Carlo techniques for describing the dynamics of complex reaction systems, and the investigation of stochastic methods for simulating catalyst particle and catalytic reactor performance. The successful advancement of such efforts requires adequate availability of supercomputer time and access to appropriate types of workstations. With the increasing complexity of the systems studied, it is also necessary to have capabilities for graphical representation of the results of computations.

CONCLUSIONS

Catalysis is a complex, interdisciplinary science. Therefore, progress toward a substantially improved vision of the chemistry and its practical application depends on parallel advances in several fields, most likely including the synthesis of new catalytic materials and recognition of the reaction path of catalytic reactions. For this reason, future research strategies should be focused on developing methods with the ability to observe the catalytic reaction steps in situ or at least the catalytic site at atomic resolution. There is also a need to link heterogeneous catalytic phenomena to the broader knowledge base in solutions and in well-defined metal complexes.

Substantial progress and scientific breakthroughs have been made in recent years in several fields including atomic resolution of metal surfaces, in situ observation of an olefin complexed to zeolite acid sites by NMR spectroscopy, and in situ characterization of several reaction intermediates by a variety of spectroscopic techniques. Theoretical modeling is ready for substantial growth as a result of progress in computer technology and theory itself. For these reasons, it is desirable to focus on areas in which the extensive scientific and technological resources of academe and industry may lead to the fastest practical results. In order of priority, these areas are

1. in situ studies of catalytic reactions;
2. characterization of catalytic sites (of actual catalysts) at atomic resolution (metals, oxides);
3. synthesis of new materials that might serve as catalysts or catalyst supports; and
4. theoretical modeling linked to experimental verification.

The development of new characterization tools, particularly in spectroscopy, has been mainly the province of academic research, and thus is expected to continue because industry is finding it increasingly difficult to justify the costs associated with technique development. To maintain the present worldwide leadership of the United States in both the science of catalysis and catalyst technology, it is essential to provide additional support for academic research focused on items 1-4 above. Furthermore, additional steps must be taken to facilitate interaction and, in fact, cooperation between industry, dealing with proprietary catalysts, and academe, developing advanced characterization tools and theory for catalysis.

4

National Resources for
Catalytic Research

The national resources for conducting research on catalysts and catalytic processes are distributed among academe, national laboratories, and industry. This chapter briefly reviews the type of research conducted in each institution and the available level of support.

INDUSTRY

The development of new chemical processes based on catalytic technology is carried out almost exclusively in industry and, to a much lesser extent, in national laboratories and universities. The commercialization of a new catalytic process is capital-intensive, and the time from the discovery of a viable catalyst for a new process to commercial plant start-up may be as long as 10 to 15 years. Research can contribute to a minimization of this time lag. Most new processes are sufficiently complex that large-scale pilot plants are required to collect the high-quality basic data needed to design and build safe, clean, and efficient commercial plants. In these pilot plants, many new issues must be resolved. The effects of operational variables, such as pressure, temperature, feed composition and purity, contact time, and recycle, on catalyst life and performance are evaluated. Often, and quite surprisingly, trace contaminants can build up and seriously affect a catalyst, so that additional improvements must be made. Physical property data must be measured, and viable methods of isolating and purifying the product and intermediates must be developed. Industry and government agencies are requiring higher standards for product quality and for disposal of waste by-products and spent catalysts. The venting of unwanted products and deep-well injection of aqueous and organic waste are receiving close attention and are likely to become unacceptable in the near future. New techniques to return these products to the process or render them innocuous must be developed. This entire development process can easily cost tens of millions of dollars and lead to a new commercial plant that may cost hundreds of millions of dollars. Multidisciplinary teams of highly trained professionals are required with expertise in areas such as kinetics; organic, inorganic, and physical chemistry; process control; materials science; separations; and all of the fundamental unit operations of mechanical and chemical engineering. The supply of talent is best provided by our educational system. Once a new process has been commercialized, very little time may be spent in providing a detailed understanding of the chemistry involved because resources are often quickly shifted to other projects to meet short-term profit objectives. Developing a basic understanding of new and significant chemical processes represents an excellent opportunity for collaboration between academic researchers and their industrial counterparts.

Industrial laboratories account for a very significant amount (greater than 90%) of the total R&D dollars spent on catalysis-related research and development. The panel conservatively estimates that the total amount of money spent annually on catalysis R&D in industry is $500 million to $1 billion. Of this, 90-95% is used for solving development and environmental problems; hence, relatively little is directed toward long-range research leading to new discoveries. The current practice of sharply restricting the investment of R&D funds in long-range research is of considerable concern, because it could adversely affect the ability of the United States to remain a world leader in the provision of new catalytic technologies.

UNIVERSITIES

A primary objective of our university system is to provide suitably educated students to enter academic, government, industrial, or other chosen careers. In addition, in the field of catalysis, research universities provide new experimental techniques, new instrumentation to study catalysts at the molecular level, new catalytic materials, and new theoretical concepts and approaches for understanding the structure of catalysts and the dynamics of catalyzed reactions. Given the significance of catalysis to the United States, an interest in science must be encouraged at all levels in the educational process to provide the necessary supply of qualified teachers and educated professionals. Students need to be educated in the basic skills of catalysis that are anticipated and required by industry. Technology is advancing at such a rapid pace that obsolete equipment must be upgraded regularly to provide students with hands-on experience in using state-of-the-art equipment. Research programs should not be closely tied to proprietary industrial needs but could easily supplement those needs. Universities must continue to provide an environment in basic science for developing new and leading-edge technologies. This is a long-range process, and so sufficient time and funding should be provided to ensure continuing progress.

Support for university research on catalysis comes primarily from the National Science Foundation (NSF), the Department of Energy (DOE), the Department of Defense (DOD), and the National Institutes of Health (NIH) and, to a much lesser extent, from the Gas Research Institute (GRI), the Electric Power Research Institute (EPRI), and the Petroleum Research Fund (PRF) of the American Chemical Society. The distribution of support is listed in Table 4.1.

During the past five to seven years, the level of federal funding in catalysis has remained constant in dollars. However, inflation and rising overhead costs have reduced the purchasing power of this support. If the United States is to retain its leadership position in the field of catalysis, its level of support for academic research in catalysis must increase. This investment will pay off not only through the provision of well-educated young talent for industrial and other research organizations, but also through a continuing expansion of the reservoir of fundamental knowledge on which all researchers depend for new ideas and concepts.

In recent years, federal support of university research has been supplemented to a small degree by industrial grants and contracts. The incentive for this support has been industry's growing awareness of global competition and the resulting pressure to control R&D expenses. These pressures have necessitated a reduction in the level of long-range, fundamental research done in industry. To offset this trend, several companies have undertaken the support of such research at universities. This support of university research by industry has led to a very good leveraging of resources. In some states there is money available, dollar for dollar, to match industrial

Table 4.1 Funding for Catalysis-Related Research at Universities (millions of dollars—1989)

Research Funded	Source of Funding							Total Funding
	NSF	DOE	DOD	NIH	GRI	EPRI	PRF	
Chemical catalysis	11.7	14.3	1.2	0.0	2.4	1.2	1.0	31.8
Biocatalysis	1.6	1.5	3.7	17.0[a]	0.0	0.1	0.0	6.9[b]

[a]Support for basic enzymology.
[b]Excluding NIH support for enzymology.
SOURCE: Data provided by staff officers of the agencies and organizations.

funding of local universities as a means of encouraging strong partnerships. In addition, for young investigators, the Presidential Young Investigators program also offers the possibility of matching funds from NSF, resulting in a potential 3:1 leveraging. It should be noted, however, that at present, the total level of industrial support for academic research is very low (3%) compared to that provided by federal agencies.

NATIONAL LABORATORIES

The Department of Energy (DOE) national laboratories represent a major resource for conducting research and development. These laboratories receive $6 billion per year from DOE and employ more than 5% of the nation's research workers. Although in the past a major fraction of the programs conducted at the national laboratories have been defense related, an increasing proportion is now devoted to energy production, environmental protection, health, and the improvement of economic competitiveness.

Research on catalysis or closely related subjects is carried out at each of the eight national laboratories. The objectives of this work encompass the development of novel instrumentation for in situ and ex situ characterization of catalysts, studies of fundamental processes occurring on catalyst surfaces, and the elucidation of catalyst-function relationships. Examples of new approaches developed by researchers at national laboratories include special high-pressure and low-pressure chambers for the characterization of practical catalysts by using surface-analytical techniques, novel optical methods exhibiting high surface specificity (e.g., second harmonic generation, sum frequency generation), and novel methods for in situ characterization of catalysts by infrared, Raman, and nuclear magnetic resonance spectroscopies.

Support by DOE through national and other laboratories has also been responsible for the design, construction, and operation of major user facilities such as the synchrotron light sources at Stanford University, Argonne National Laboratory, Brookhaven National Laboratory, and Lawrence Berkeley National Laboratory; the pulsed neutron source at Argonne National Laboratory; and the National Electron Microscopy Center at Lawrence Berkeley National Laboratory. Use of these facilities is available to researchers from academe and industry, as well as those working at the national laboratories themselves.

At present, DOE provides about $14 million annually to support catalysis research at the national laboratories. This constitutes approximately 50% of the total DOE support for catalysis research. It should be noted that at the Ames National Laboratory and the Lawrence Berkeley National Laboratory, which have close ties to Iowa State University and to the University of California at Berkeley, respectively, most of the research on catalysis is carried out by graduate students working under the direction of a faculty member.

Because of the unique capabilities available at the national laboratories, several companies have found it attractive to sponsor research at these facilities and to send their employees to national laboratories for training in novel or highly specialized techniques. These relationships have contributed to an increasing level of technology transfer from the national laboratories to industry. Technology transfer has also occurred through the employment of students and postdoctoral fellows who received their professional training while working at a national laboratory.

Although issues of patents and patent rights have been an impediment to industrial support of research at national laboratories in the past, recent years have witnessed the development of more flexible contractual agreements. Such agreements have been successful in protecting proprietary information provided by the company and in offering opportunities for the company to seek patents in its own name, based on the research it supported.

5

Findings and Recommendations

The present study demonstrates that catalysis is a critical technology underlying two of the largest industries in the United States, the chemical and petroleum processing industries, and is a vital component of a number of the national critical technologies identified recently by the 1991 *Report of the National Critical Technologies Panel*. Catalysis is also shown to be essential for most modern, cost- and energy-efficient means of environmental protection and for the production of a broad range of pharmaceuticals. The impact of catalysis on the nation's economy is clearly evident from the fact that catalytic technologies generate sales in excess of $400 billion per year and a net positive balance of trade of $16 billion annually.

Although the chemical industry in the United States remains strong, it faces increasing competitive challenges from the European Community and Japan, and in fact, the three largest chemical companies are no longer based in the United States but rather in Germany. Growing recognition that environmental protection must be achieved at all stages of the production, use, and disposal of chemicals represents a second major challenge. Added to this is the continuing challenge to find economically efficient, and environmentally acceptable, means of producing transportation and heating fuels from petroleum and other fuel resources.

As detailed in this report, catalysis will play a vital role in addressing each of the three challenges cited above. The development of low-cost, environmentally benign methods for producing chemicals requires the discovery and development of new catalysts. The reduction of toxic emissions from stationary sources requires the development of cost-efficient catalytic processes. The production of polymers and pharmaceuticals, molecularly designed to achieve specific applications, will depend on new catalysts.

70

The panel has also shown that through the use of modern analytical and theoretical techniques, scientific investigation of catalysis has led to an unprecedented understanding of catalysts and catalytic processes at the molecular level. These investigations have opened the door to a detailed understanding of structure-function relationships and of the effects of reaction conditions on the structure and composition of catalysts. The advent of supercomputers and improved theoretical methods is rapidly enabling the simulation of many aspects of catalysis. Taken together, these advances have contributed to making catalysis less an art and more a science, and have assisted and accelerated the process of developing new catalysts for industrial applications.

The strong position in catalytic science and technology held by the United States is the result of past investments made by both industry and the government. Recent years have seen a decrease in the rate of investment by industry—particularly in ongoing, long-range, fundamental and exploratory research programs—as a consequence of competitive pressures, corporate mergers, and placement of resources in business ventures outside the central activities of the industry. Concurrently, government has maintained a relatively constant dollar investment in catalysis, but inflation and rising overhead costs have eroded the purchasing value of these funds. The net result is that the United States has lost momentum just at the moment that it must face a significant number of challenges and opportunities. *To take advantage of these opportunities, careful attention must be given to effective use of the nation's resources, so that the United States can maintain its leadership role.*

The balance of this chapter presents specific recommendations for action by industry, academe, the national laboratories, and the federal government.

INDUSTRY

As detailed in Chapter 2, substantial opportunities exist for developing new processes and products, pending the development of as yet unavailable catalytic technologies. It is important to note that these new, economically favorable catalytic processes, once developed, will be adopted globally, providing long-range economic benefits to the originating company and country. Given these opportunities, combined with the challenges of utilizing new raw materials (e.g., methane and coal) and protecting the environment, industry should strive to readjust the balance internally or externally between long- and short-range research. Internally, this would be facilitated by long-range business and technology planning, technology forecasting and trend analysis, a more stable commitment to strategic projects, joint development and joint venture programs with other companies for risk sharing, and high-quality project selection and evaluation methodologies.

Turning to external ways for industry to redress the balance between short- and long-term research, the panel notes that many of the challenges faced by industry will require additional advances in the science of catalysis, as well as advances in instrumentation. Given the current cost of conducting research in industry, opportunities exist for developing meaningful collaborative programs in partnership with academic and national laboratory researchers. To achieve this goal two elements are recommended as essential:

1. Enhanced appreciation by academic researchers of industrial technology. Vehicles for this include
   • **long-term consulting arrangements involving regular interactions with industrial researchers,**
   • **sabbaticals for industrial scientists in academic or government laboratories,**
   • **sabbaticals for academic or government scientists in industrial laboratories,**
   • **industrial internships for students,**
   • **industrial postdoctoral programs, and**
   • **jointly organized symposia on topics of industrial interest.**

2. Increased industrial support of research at universities and national laboratories. Vehicles for this include
   • **research grants and contracts;**
   • **unrestricted grants for support of new, high-risk initiatives; and**
   • **leveraged funding (e.g., support of the Presidential Young Investigators program.)**

ACADEMIC RESEARCHERS

Over the past 25 years, academic researchers have made major contributions to understanding the structure of catalysts and the relationships between structure and function. These efforts have also resulted in the development of new instrumental and theoretical techniques, many of which now find application in industrial laboratories. As discussed in Chapter 3, progress in catalyst science must be sustained to provide the basis for future developments in catalyst technology and for the continuing supply of men and women educated in the scientific principles that underlie catalysis. The panel, therefore, makes the following recommendations to the academic community:

1. A materials-focused approach is needed to complement the existing strong efforts on understanding and elucidating cata-

lytic phenomena. **More emphasis should be placed on investigation of the optimized design and synthesis of new catalytic materials, in addition to the study of existing ones. It must be kept in mind that a new material deserves consideration as a potential catalytic material only after its successful use as a catalyst, or as a component of such.**

2. Further advancement should be made in the characterization of catalysts and the elucidation of catalytic processes, particularly under reaction conditions; existing studies of structure-function relationships should be continued and expanded to focus on catalysts relevant to applications with major potential.

3. Academic researchers should develop cooperative, interdisciplinary projects, or instrumental facilities, in which researchers from a range of disciplines work on various aspects of a common goal, as exemplified by programs carried out in NSF-supported Science and Technology Centers.

4. Academic researchers should be encouraged to work collaboratively on projects with industry that are aimed at enabling the development of catalyst technology through the application of basic knowledge of catalysts and catalytic phenomena.

5. Academic institutions should ease their patent policies with respect to ownership and royalties, to facilitate greater industrial support of research.

NATIONAL LABORATORIES

National laboratories have been organized around large-scale, national issues requiring high-technology, focused, team-oriented work, often combined with the ability to take major risks in an economically shielded environment. These laboratories are truly national resources. They have been highly effective in developing novel instrumentation for catalyst characterization, operating large-scale user facilities (i.e., synchrotron radiation sources, pulsed neutron sources, and atomic resolution microscopes), and applying the most advanced experimental and theoretical techniques to study structure-function relationships critical for understanding catalysis at the molecular level. Given the wealth of resources at the national laboratories, major opportunities exist for advancing catalyst science and technology through research carried out in collaboration with industry and academe. To achieve this goal, national laboratories are encouraged to

1. undertake joint research projects with industry focused on developing a fundamental understanding of the structure-property relationships of industrially relevant catalysts and

catalytic processes, and on using such understanding for the design of new catalysts for major new process opportunities;

2. continue the development of novel instrumentation for in situ studies of catalysts and catalytic phenomena;

3. place greater emphasis on the systematic synthesis of new classes of materials of potential interest as catalysts; and

4. investigate novel catalytic approaches to the production of energy (e.g., light-assisted catalytic splitting of water), the selective synthesis of commodity and fine chemicals, and the protection of the environment.

FEDERAL GOVERNMENT

The principal sources of funding for university and national laboratory research on catalysis are the Department of Energy (DOE) and the National Science Foundation (NSF). As noted in Chapter 4, constant-dollar funding from these agencies, together with inflation and rising overhead costs, has caused a decrease in the number of young scientists being educated in the field of catalysis. The panel also observes that with the decline in emphasis on alternative fuels, research in catalysis has become increasingly diversified and less aligned along national interests. To offset these trends, the panel recommends that federal agencies:

1. establish mechanisms for reviewing their programs related to catalysis, to ensure that they are balanced and responsive to the needs of the nation and to the opportunities for accelerating progress;

2. encourage industry to assist the funding agencies in identifying important fundamental problems that must be solved to facilitate the translation of new discoveries into viable products and processes; assessment of the fundamental research needs of industry should be communicated to all members of the catalysis community; and

3. increase the level of federal funding in support of catalysis research by at least a factor of two (after correction for inflation) over the next five years. This recommendation is consistent with the Bush administration's proposal to double the NSF budget over the next five years and with a recent statement by Frank Press, president of the National Academy of Sciences, that doubling the research budgets of all federal agencies should be a goal for the 1990s. Recognizing the need for federal agencies to maintain flexibility and to encourage creative scientists who propose to explore new directions and ideas, the panel recommends that priority be given to the following five areas:

- Synthesis of new catalytic materials and understanding of the relationships between synthesis and catalyst activity, selectivity, and durability.
- Development of in situ methods for characterizing the composition and structure of catalysts, and structure-function relationships for catalysts and catalytic processes of existing, and potential, industrial interest.
- Development and application of theoretical methods for predicting the structure and stability of catalysts, as well as the energetics and dynamics of elementary processes occurring during catalysis, and use of this information for the design of novel catalytic cycles and catalytic materials and structures.
- Investigation of novel catalytic approaches for the production of chemicals and fuels in an environmentally benign fashion, the production of fuels from non-petroleum sources, the catalytic abatement of toxic emissions, and the selective synthesis of enantiomerically pure products.
- Provision of the instrumentation, computational resources, and infrastructure needed to ensure the cost-effectiveness of the entire research portfolio.

Appendix

To gather information for its report, the panel contacted a large number of individuals in both academe and industry. Each was asked to respond to the following questions:

- What areas of fundamental research are most helpful to support commercial catalyst/catalysis activity in U.S. industry?
- Should the dispersal of federal research grants to academic researchers be based on demonstrated excellence in science or focused to support the national laboratories?
- What type of linkage with academia/national laboratories is most useful to, and supportable by, U.S. industry?
- What elements in science or technology provided the edge to your commercial business in catalyst/catalytic processes?
- What novel catalytic processes do you expect to be developed in the next 10 to 15 years?
- What will be the nature of the exploratory and basic research that leads to these developments?
- Is academic and industrial catalytic research in the United States well positioned to play a leadership role in creating this new technology and, if not, what needs to be done?
- Identify areas of catalyst science and technology in which the United State is (1) behind competitors, (2) even with competitors, and (3) ahead of competitors.
- Identify problems that have long-term payoff.

• What areas are "mature" or "dead"?

• Has too much emphasis been placed on one area in the past?

• What would be the ideal mix of industrial and academic research in catalysis?

• What are the major unsolved problems in catalysis, and what would the solution to these problems provide in economic and technical terms?

• Are there new areas where catalysis could be used?

A total of 30 responses to these questions was received. Those providing input are acknowledged below as corresponding contributers.

CORRESPONDING CONTRIBUTORS

Charles R. Adams
Shell Development Company

David Allen
Department of Chemical Engineering
University of California,
  Los Angeles

Paul A. Bartlett
Department of Chemistry
University of California, Berkeley

Jay B. Benziger
Departmnet of Chemical Engineering
Princeton University

Robert G. Bergman
Department of Chemistry
University of California, Berkeley

Cynthia J. Burrows
Department of Chemistry
State University of New York,
  Stony Brook

James P. Collman
Department of Chemistry
Stanford University

Mark E. Davis
Department of Chemical Engineering
California Institute of Technology

W. Nicholas Delgass
Department of Chemical Engineering
Purdue University

Francois N. Diederich
Department of Chemistry
  and Biochemistry
University of California,
  Los Angeles

Robert P. Eischens
Zettlemoyer Center for Surface
  Science
Lehigh University

John G. Ekerdt
Department of Chemical Engineering
University of Texas, Austin

David A. Evans
Department of Chemistry
Harvard University

Rocco A. Fiato
Exxon Research and Engineering
  Company

Juan M. Garces
Dow Chemical Company

Mary L. Good
Signal Research

Vladimir Haensel
Chemical Engineering Department
University of Massachusetts,
  Amherst

Gary L. Haller
Chemical Engineering Department
Yale University

Heinz Heinemann
Center for Advanced Materials
Lawrence Berkeley Laboratory

Enrique Iglesia
Exxon Research and Engineering
  Company

William P. Jencks
Graduate Department of Bio-
  chemistry
Brandeis University

Andrew S. Kaldor
Exxon Research and Engineering
  Company

Jeremy R. Knowles
Department of Chemistry
Harvard University

Ralph Landau
Listowel, Inc.

Jerry A. Meyer
Chevron Research and Technology
  Company

Craig B. Murchison
Dow Chemical Company

Mario L. Occelli
Unocal Corporation

Nicholas D. Spencer
W. R. Grace & Company

George M. Whitesides
Department of Chemistry
Harvard University

Craig Wilcox
Department of Chemistry
University of Pittsburgh

As an additional means of gathering information, the panel held a workshop on April 20-21, 1990, to which it invited a series of speakers to give a perspective on the current status of catalysis research and prospective areas for future work. Representatives from each of the federal agencies supporting catalysis research were invited to present a summary of their programs.

PROGRAM FOR THE WORKSHOP ON NEW DIRECTIONS IN CATALYST SCIENCE AND TECHNOLOGY

Friday, April 20, 1990

8:30-8:45 a.m.	Alexis Bell—University of California, Berkeley Introduction and Overview
8:45-9:25 a.m.	James Cusumano—Catalytica, Inc. Catalytic Technologies
9:25-9:45 a.m.	Discussion
9:45-10:25 a.m.	N. Y. Chen—Mobil Research and Development Company Zeolite Catalysis
10:25-10:45 a.m.	Discussion
10:45-11:00 a.m.	Break
11:00-11:40 a.m.	James Lyons—Sun Oil Company Alkane Activation by Partial Oxidation
11:40-12:00 a.m.	Discussion
12:00-1:00 p.m.	Lunch
1:00-1:40 p.m.	Mordecai Shelef—Ford Motor Company Catalysis for Environmental Protection
1:40-2:00 p.m.	Discussion
2:00-2:40 p.m.	George Parshall—E. I. Du Pont de Nemours & Company Industrial Synthesis of Chemicals via Homogeneous Catalysis
2:40-3:00 p.m.	Discussion
3:00-3:15 p.m.	Break
3:15-3:55 p.m.	Jack Halpern—University of Chicago New Directions in Homogeneous Catalysis
3:55-4:15 p.m.	Discussion
4:15-4:55 p.m.	Fred Karol—Union Carbide Corporation Polymerization Catalysis
4:55-5:15 p.m.	Discussion

Saturday, April 21, 1990

8:30-9:10 a.m.	John Tully—AT&T Bell Laboratories
	Theory Applied to Gas-Surface Interactions
9:10-9:30 a.m.	Discussion
9:30-10:10 a.m.	Jack Kirsch—University of California, Berkeley
	New Challenges in Biocatalysis
10:10-10:30 a.m.	Discussion
10:30-10:45 a.m.	Break
10:45-11:25 a.m.	Dennis Forster—Monsanto Company
	The Interface of Catalysis with Biology
11:25-11:45 a.m.	Discussion
12:00-1:00 p.m.	Lunch
1:00-1:15 p.m.	Robert Mariannelli—U.S. Department of Energy
1:15-1:30 p.m.	Kendall Houck—National Science Foundation
1:30-1:45 p.m.	Warren Jones—National Institutes of Health
1:45-2:00 p.m.	Harold Guard—Office of Naval Research
2:00-2:20 p.m.	Discussion
2:20-3:30 p.m.	General Discussion

Index